SO-CBU-085

SpringerBriefs in Earth System Sciences

Series Editors

Gerrit Lohmann
Lawrence A. Mysak
Justus Notholt
Jorge Rabassa
Vikram Unnithan

For further volumes:
http://www.springer.com/series/10032

Gerrit Lohmann · Klaus Grosfeld
Dieter Wolf-Gladrow · Vikram Unnithan
Justus Notholt · Anna Wegner
Editors

Earth System Science: Bridging the Gaps between Disciplines

Perspectives from a Multi-Disciplinary Helmholtz Research School

 Springer

Editors
Prof. Dr. Gerrit Lohmann
Alfred Wegener Institute for Polar and
    Marine Research
Bremerhaven
Germany

Dr. Klaus Grosfeld
Alfred Wegener Institute for Polar and
    Marine Research
Bremerhaven
Germany

Prof. Dr. Dieter Wolf-Gladrow
Alfred Wegener Institute for Polar and
    Marine Research
Bremerhaven
Germany

Prof. Dr. Vikram Unnithan
School of Engineering and Science
Jacobs University Bremen gGmbH
Bremen
Germany

Prof. Dr. Justus Notholt
Institute of Environmental Physics
University of Bremen
Bremen
Germany

Dr. Anna Wegner
Alfred Wegener Institute for Polar and
    Marine Research
Bremerhaven
Germany

ISSN 2191-589X   ISSN 2191-5903   (electronic)
ISBN 978-3-642-32234-1   ISBN 978-3-642-32235-8   (eBook)
DOI 10.1007/978-3-642-32235-8
Springer Heidelberg New York Dordrecht London

Library of Congress Control Number: 2012944976

Preface

Promoting young researchers is a major priority of the Helmholtz Association, Germany's largest scientific research organisation with 18 Research Centres in the fields of science, engineering and biomedicine. In the framework of the Initiative and Networking Fund several promotion instruments were set up with the aim to advance education and to attract excellent young people for science conducted by the Helmholtz Centres. School laboratories, doctoral training programmes, as well as Postdoc programmes support individuals at every stage of their education and career. Helmholtz Research Schools aim to prepare highly skilled doctoral students for a career in science and business. Each research school brings together up to 25 outstanding young doctoral students to conduct research on a specific topic and thus gain valuable experience working together closely in teams—an absolutely essential skill for topnotch research today. In addition, the Helmholtz Association works with distinguished partners such as the Imperial College London, enabling it to provide a curriculum that includes a range of courses that aim to foster professional qualification and personal development and to equip graduates for careers in management positions, both in science and the business world. Since 2006, 21 research schools have been set up at the different Helmholtz Centres. The research schools are complemented by 13 Helmholtz Graduate Schools which provide a roof for a varied number of curricula in different fields, or across disciplines.

At the Alfred Wegener Institute for Polar and Marine Research in Bremerhaven, a Helmholtz Research School on Earth System Sciences has been funded since 2008 in collaboration with the University of Bremen and the Jacobs University Bremen. Using the network and collaboration of experts and specialists from the different institutes on observational and paleo-climate data as well as on statistical data analysis and climate modelling, doctoral students from eight countries were trained to understand, decipher and cope with the challenges of global climate change. The Earth System Science Research School (ESSReS) covers all kinds of disciplines, climate science, geosciences and biosciences, and provides a consistent framework for education and qualification of a new generation of expertly trained, internationally competitive doctoral students.

The set-up of a structured doctoral programme like ESSReS combines both, strong scientific cutting-edge research and an interdisciplinary education that bridges the gap between the traditional disciplines. The young students are motivated to learn on an interdisciplinary and trans-institutional basis, guiding their way in modern research. The success and outcome of the first 3 years phase of ESSReS, which also served as precursor for the Graduate School for Polar and Marine Research (POLMAR), established at the Alfred Wegener Institute in 2009, is visible in this book. Both schools provide a new level with binding rules for doctoral education at the Alfred Wegener Institute, satisfying our enduring efforts on the improvement of doctoral education in the Helmholtz Association.

Bremerhaven, September 2012   Prof. Dr. Karin Lochte
                                                               Prof. Dr. Jürgen Mlynek

Acknowledgments

Funding by the Helmholtz Association and continuous support of the Earth System Science Research School (ESSReS) of the Alfred Wegener Institute for Polar and Marine Research, the University of Bremen and the Jacobs University of Bremen is gratefully acknowledged. The editors wish to thank Dr. Peter Köhler, Dr. Martin Werner, Dr. Gregor Knorr, Andrea Bleyer, Renate Kuchta, and Stefanie Klebe (all AWI) for their kind support during the planning and implementation phases of ESSReS and the review and editing processes of this book.

Contents

Chapter 1
Introduction

1.1 General Aspects of Earth System Science

Gerrit Lohmann[1] (✉), **Klaus Grosfeld**[1], **Dieter Wolf-Gladrow**[1],
Vikram Unnithan[3], **Justus Notholt**[2] **and Anna Wegner**[1]

[1]Alfred Wegener Institute for Polar and Marine Research Bremerhaven, Germany
e-mail: Gerrit.Lohmann@awi.de
[2]Institute of Environmental Physics, University of Bremen, Germany
[3]Jacobs University, Bremen, Germany

To properly address the pressing question of climate change and its natural and anthropogenic causes, intimate knowledge on amplitude and rapidness in the natural variations of temperature or other temperature-related environmental properties in the ocean, over the continents, and in the cryosphere is required.

The best way to gain this knowledge is the inspection of historical time series of direct temperature measurements or documentation of such environmental observations. Unfortunately, historical records of direct temperature measurements which would allow consideration of changing climate on a global scale are too short and fall already within the period of strong human impact on natural conditions. Information on earlier times can be obtained either from proxies that record past climate and environmental conditions, or by simulating climate using comprehensive models of the climate system under appropriate external forcing.

One of the greatest challenges is getting reliable assertions regarding future global climate and environmental change. The longer the time scale the more components of the Earth system are involved, e.g. weather prediction models can take the ocean as constant in order to estimate the next days; interannual to decadal variations can be described in the coupled atmosphere–ocean-sea ice system, whereas longer variations like glacial-interglacial transitions incorporate the full carbon cycle as well the ice sheets and associated feedbacks (Fig. 1.1). In order to

G. Lohmann et al. (eds.), *Earth System Science: Bridging the Gaps between Disciplines*, SpringerBriefs in Earth System Sciences,
DOI: 10.1007/978-3-642-32235-8_1, © The Author(s) 2013

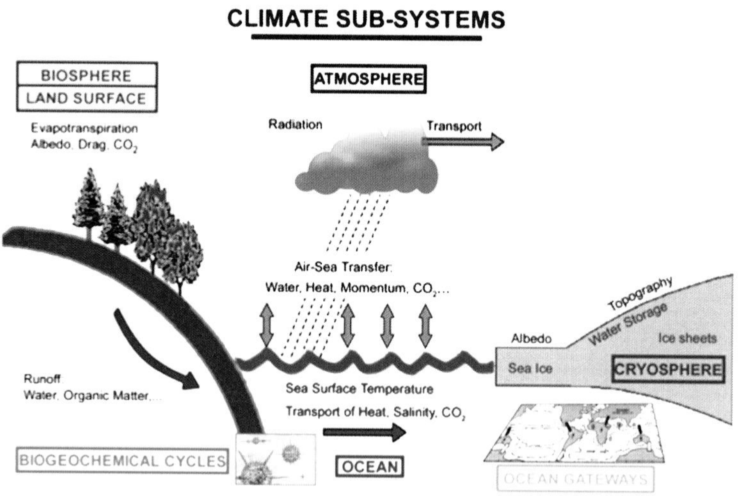

CLIMATE SUB-SYSTEMS

Fig. 1.1 The complex climate system requires a multi-disciplinary approach. The figure shows the interactions between different components. The biogeochemical cycles (including foraminifera) are a key element for the carbon cycle in the oceans and on land, feeding back to the atmosphere through the CO_2 effect on long wave radiation. The position and extent of the big ice sheets are important for the global climate system. The long-term climate cooling is also related to the carbon cycle and ocean gateway configurations, known as tectonic-sedimentary puzzle

get a systematic view of the Earth System, one has to consider the climate components with its specific time scales.

Earth System Science is traditionally split into various disciplines (Geology, Physics, Meteorology, Oceanography, Biology etc.) and several sub-disciplines. Overall, the diversity of expertise provides a solid base for interdisciplinary research. However, gaining holistic insights of the Earth System requires the integration of observations, paleo-climate data, analysis tools and modelling. These different approaches of Earth System Science are rooted in different disciplines that cut across a broad range of timescales. It is, therefore, necessary to link these disciplines at a relatively early stage in PhD programs. The linking of 'data and modelling', as a special emphasis in our graduate school, enables graduate students from a variety of disciplines to cooperate and exchange views on the common theme of Earth System Science, which leads to a better understanding of processes within a global context.

A conceptual unification among the sciences of the Earth has never developed in the German education system. Disciplinary specialization has played a large role instead of integration. Already Humboldt proposed an integration of several disciplines, establishing international cooperative networks of meteorological and geomagnetic measurement stations. Humboldt's general physics of the Earth envisioned climate as a major control of Earth-surface phenomena.

The modern view takes the Earth's land surface, oceans, atmosphere, and inhabitants as an integrated whole, with linkages among the various components (Fig. 1.1). This effort is often referred to as Earth System Science. It provides a challenge for theory, observations, reconstructions and modelling, to describe past, present and future changes in the Earth system. Global environmental change is probably one of the greatest challenges faced by human societies ever since. As a logical step, scientists need to understand the interactions and feedbacks among the components of the Earth system, encompassing the lithosphere, atmosphere, hydrosphere, cryosphere, and biosphere.

Basic knowledge in the other disciplines in Earth System Science enables graduate students from a variety of disciplines to cooperate and exchange views on the common theme of global environmental change. In our approach, we tried to cover a wide spectrum of Earth System Science: Remote sensing, data exploration, process understanding, modelling, and informatics. Informatics is also relevant since a large amount of data is retrieved through models, satellites and high-resolution geosciences data. Examples of how to link the different disciplines as a key concept of future PhD education in Earth System Science are shown in the chapters of the book. The different research projects are clustered in respect to their common research field, such as remote sensing and modelling of atmospheric chemistry, Earth system modelling and data analysis, geo-tectonics, climate archives, ecosystems and climate change, geo-informatics and geo-engineering.

1.2 The Structural and Educational Concept in an Interdisciplinary Research School for Earth System Science

Klaus Grosfeld[1] (✉), **Gerrit Lohmann**[1], **Dieter Wolf-Gladrow**[1], **Annette Ladstätter-Weißenmayer**[2], **Justus Notholt**[2], **Vikram Unnithan**[3] and **Anna Wegner**[1]

[1]Alfred Wegener Institute for Polar and Marine Sciences Bremerhaven, Germany
e-mail: Klaus.Grosfeld@awi.de
[2]Institute of Environmental Physics, University of Bremen, Germany
[3]Jacobs University, Bremen, Germany

Today, Earth System Sciences are based on highly interdisciplinary research, demanding a broad basic knowledge of the different research aspects, their feedbacks and interconnections. A major difficulty in cross-disciplinary science is, therefore, to find a proper common language/level, where people from different disciplines can explain their research question and contribute to solution strategies. Consequently, the major goal of the Earth System Science Research School (ESSReS) is bridging theses gaps by providing a scientific education and basic and expert knowledge on Earth System Science, which has not been given within the master or diploma studies of the applicants that are specialized on their research fields. By these efforts, the PhD students obtain systematic insights into the different disciplines in order to be able to

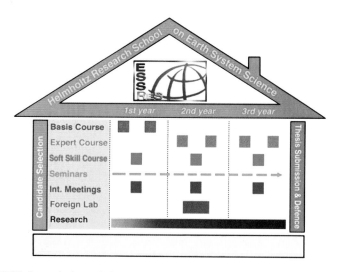

Fig. 1.2 ESSReS curriculum during the 3-years PhD education. Basic courses provide a common frame and scientific language. Transferable skills courses are related to social competence, management skills, oral and poster presentation techniques and scientific writing, as well as career planning

discuss common research questions on a solid fundament and to identify cooperation and collaboration across disciplines.

Hence, a structured educational PhD programme has been developed, which consists of basic and expert courses in the field of Earth System Science, gaining knowledge and tools, all the way to brand new developments and scientific topics (Fig. 1.2).

ESSReS is designed as a class of up to 25 PhD students from all around the globe starting the educational programme together at a certain date and following the programme as a cohort. We preferred this concept for our research school against the more common ones, where new PhDs are recruited continuously and where the research school provides a framework for different graduation stages. The advantage of our concept is that from the beginning, the PhD students build a close scientific and social network.

In total, two weeks of lectures during the introductory phase and one week of lectures per semester during the expert phase are taught. The lectures are complemented by monthly PhD seminars that are organized by the PhD students themselves and hosted in turns at the different institutes (Fig. 1.3).

Once a year, the PhD students present their results with oral and poster presentations at a common retreat to their colleagues and the supervisor assembly, giving room to extensive discussion and interdisciplinary exchange. For expert courses external lecturers are invited for an extended research stay at one of the collaborating institutes, to teach challenging new topics relevant for the research school and to be available during their research stay for intensive discussions and collaborations with

Fig. 1.3 Monthly ESSReS seminar with self-organized presentations and discussion about the different research topics and planning of common activities

the PhD students. An overview on the curriculum and course programme can be found on our research School web page www.earth-system-science.org. Additional courses, interesting for the PhD students and provided by our cooperation graduate schools at the Alfred Wegener Institute (www.polmar.awi.de), at the Bremen University (www.glomar.de) or at the Max Planck Institute Research School for Marine Microbiology (www.marmic.mpg.de) are accepted within the curriculum.

The scientific education is complemented by a three-stage training in transferable skills. Besides the scientific work, the PhD students join each year a residential course organized by the Helmholtz Association and performed by tutors from the Imperial College London, where "Research Skill Development", "Presentation and Communication Skills", and "Career Planning and Leadership Skills" are trained. The courses are run together with participants from other Helmholtz research Schools to foster further interdisciplinary experience and networking within the young researchers' community of the Helmholtz Association. In addition to the transferrable skills training, a special coaching programme for scientific writing is offered to the PhD students through the foreign language institute at the University of Bremen, to support the writing of their first scientific papers directly from the beginning. Deficiencies in scientific writing have shown to be a great obstacle in the time planning, because the first paper takes empirically much longer than expected.

During the programme, students will be required to directly apply their soft skills. Fellows may wish to receive financial assistance for activities beyond what is generally covered by the research school, *e.g.* for additional travel to conferences, summer schools or lab support. To this end, the research school provides a certain budget for each PhD student. To access these funds, fellows are required to write short proposals that are evaluated by the Academic Council. In writing these proposals and reports, the PhD students undergo basic training in fundraising and fund management which is an important part of professional research right from the beginning. Furthermore, this represents a good exercise for applicants who intend to apply for a research grant for an extended internship at a foreign lab, which is highly recommended by ESSReS.

1.2.1 The Supervision and Mentoring Concept

Each student's PhD research is supervised by a PhD committee, which consists of the supervisor and two senior scientists. For interdisciplinary Earth System Sciences it proved useful to complement the committee by an expert from another research field. The PhD student should be encouraged to add further persons to their committee if necessary or desirable (e.g., another PhD student). The PhD students meet with their committee at least twice a year to discuss their efforts and monitor their development. The first meeting and initiation of the committee takes place no later than 6 months after the start of the PhD project where a proposal of the PhD project is presented and discussed, including title, aim, literature review and work plan, to arrange a guideline for the 3-years PhD work. For each meeting a short report is written, documenting the research strategy, plans and aims for the next 6 months. This builds the base for the beginning of the next meeting, where a review of the last 6 months shows problems, progress and results of the PhD work. The PhD meeting is a mutual opportunity for PhD student and supervisor to discuss and plan the PhD progress on a regular basis for getting an early hand on the time management of the whole project.

1.2.2 The Managing Concept

ESSReS is organized by a clear administrative and executive structure which is led by the Advisory Board, consisting of the (di)rectors of all participating institutions (Fig. 1.4). It assumes the ultimate authority over the research school. The Advisory Board is responsible for strategic decisions concerning the research school. The Academic Council is the executive board of ESSReS. It is composed of the speaker of the research school who represents the research school, one representative from each of the participating institutions, the coordinator of the research school, and a PhD student and supervisor delegate, respectively. The responsibilities of the Academic Council are to oversee and develop all academic matters, including the development of scientific fields, outward scientific presentation, invitation of guest scientists, student recruitment and admission, academic integrity, conflict solution, and review of funding proposals. The Academic Council is supported by the research school coordinator, who represents the research school vis-à-vis the participating institutions. The coordinator manages and administrates all matters that affect the research school, its cooperation with the three institutions' administrations, and liaison with the funding institutions. The position of the coordinator is of central importance for the success of the research school, because

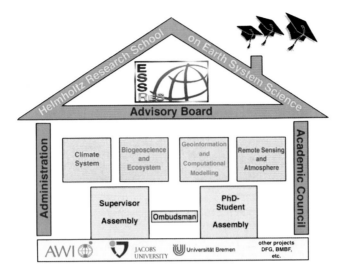

Fig. 1.4 ESSReS management structure and main scientific topics

the coordinator is responsible for the organization of the research school and course planning, workshops and further activities, and is the direct contact person for the PhD students in all concerns. Furthermore, the coordinator takes care of the administrative aspects of the research school and is the direct contact person for the students, the administration, the Student Assembly, the Supervisors Assembly, and the Academic Council. It is advantageous for the coordinator to be a scientist with a background in a broad field of Earth System Science and an active partner in the research and educational programme.

Two additional boards are of central importance for the management of the research school, the Supervisor Assembly and the Student Assembly. All major decisions regarding the research school's educational and PhD research programme are taken after thorough debate within the Supervisor Assembly. This assembly also proposes guest scientists for advanced lectures and suggests new topics in the educational programme. The Supervisor Assembly is chaired by an elected person, who is delegated for the Academic Council. The Student Assembly comprises all PhD students of the research school. It works out its own rules and procedures, autonomously. The assembly is the forum for the PhD students to discuss all issues related to the graduate training they receive at the research school and its related research institutions. Here, demands in terms of course topics and supervision requirements as well as deficiencies in the programme can be formulated and communicated to the Academic Council. Both assemblies convene at least once a year during the annual retreat and the representatives participate actively in the Academic Council.

The coordinator strives to solve arising problems of any kind. Otherwise, an Ombudsman is available to moderate discussions between the affected parties. This helps to clarify any problems which may avert the success of the PhD thesis.

1.2.3 The Helmholtz Certificate

The Helmholtz Association has developed a standardized certificate for all its Research and Graduate Programmes in order to document and certify the given education. The Helmholtz certificate describes the general outline of the research school and the collaborating institutions, and documents all courses, seminars, summer schools, active conference participations and expeditions of the candidate in a transcript. The certificate is given at the end of the 3-years training programme, after fulfilling the minimum of 45 credit points in 3 years, where each credit point is comparable to one working day. The certificate is a testimony of the attended courses in the field of Earth System Sciences beyond the specific expertise gained through the PhD thesis and shall serve as proof for the additionally gained qualification.

Chapter 2
Remote Sensing and Modelling of Atmospheric Chemistry and Sea Ice Parameters

2.1 NO$_2$ Pollution Trends Over Megacities 1996–2010 from Combined Multiple Satellite Data Sets

Andreas Hilboll (✉), Andreas Richter and John P. Burrows

Institute of Environmental Physics, University of Bremen, Germany
e-mail: hilboll@iup.physics.uni-bremen.de

Abstract Nitrogen oxides (NO$_x$ = NO$_2$ + NO) are air pollutants emerging mainly from fossil fuel combustion, i.e. traffic, power generation, and industry. Apart from being hazardous to human health, they contribute to acid rain and play a major role in tropospheric ozone formation. While NO$_x$ concentrations can most accurately be monitored using ground-based in-situ measurements, remote sensing techniques in general and satellite instruments in particular have proven invaluable to obtain long and consistent time series with global coverage. These data sets facilitate studying the temporal evolution of atmospheric pollutants like NO$_2$, as they allow applying identical measurement techniques to all investigated regions, yielding comparable results. In this study, we present an assessment of the evolution of tropospheric NO$_2$ for the 1996–2010 time period. Satellite measurements from the GOME, SCIAMACHY, OMI, and GOME-2 instruments are used in an ensemble approach to investigate trends in tropospheric NO$_2$. The focus is on large urban agglomerations, where air quality is most important for human health. Our findings show generally decreasing NO$_2$ levels for most urban agglomerations of the developed world, while annual growth rates for developing cities can be as high as 9 % annually.

Keywords Air pollution · Megacities · Trend analysis · Nitrogen dioxide · GOME · SCIAMACHY

G. Lohmann et al. (eds.), *Earth System Science: Bridging the Gaps between Disciplines*, SpringerBriefs in Earth System Sciences,
DOI: 10.1007/978-3-642-32235-8_2, © The Author(s) 2013

2.1.1 Introduction

Since the year 2009, more than half of the Earth's human population is living in cities (United Nations Department of Economic and Social Affairs 2010). This trend of increasing urbanization is particularly pronounced in the developing countries, especially in megacities of 10 million inhabitants and more. Due to the very high population density, energy use, and traffic, these agglomerations are hot-spots in terms of air pollution, having large negative health impacts on an increasing number of people (Smith 1993).

One of the main atmospheric pollutants in a megacity environment is the class of nitrogen oxides ($NO_x = NO + NO_2$). NO_x are formed as NO by the reaction of nitrogen and oxygen atoms, disassociated from their molecular state by the high temperatures found in industrial and traffic combustion processes.[1] In daylight conditions, NO and NO_2 then undergo constant cycling (Seinfeld and Pandis 2006). Due to its relatively short lifetime of several hours to a few days, the natural sources of nitrogen oxides (biomass burning, lightning, soils) account for most tropospheric NOx in rural areas. They are, however, not relevant in most megacity environments. While NO_2 itself is very poisonous, its impact on human health and the environment is even further augmented by its major role in the formation of tropospheric ozone and, via the formation of nitric acid, as acid rain (Seinfeld and Pandis 2006).

While ground-based in situ instruments have been employed to monitor NO_x concentrations for a long time, NO_2, among other trace gases, can also be investigated using remote-sensing techniques. As light travels through the atmosphere, it is partly absorbed by trace constituents along the way following the Beer-Lambert law, which allows for using spectroscopic approaches. One example is differential optical absorption spectroscopy (DOAS) (Platt and Stutz 2008), which yields the integrated number of absorber molecules along the average light path through the atmosphere, called the total slant column density. In recent years, such DOAS instruments have been mounted on satellite platforms, granting the advantage of comparable measurement conditions for all locations, global coverage every few days, and reduced biases introduced by the location of possible ground-based measurement stations. On the other hand, satellite measurements suffer from low spatial resolution, often cloudy scenes, and quite high uncertainties, which are due to non-perfect knowledge of the exact light path through the atmosphere.

Since 1995, a total of four satellite-based instruments provides measurements of tropospheric NO_2 by means of DOAS: GOME (Burrows et al. 1999), SCIAMACHY (Bovensmann et al. 1999a, b), OMI (Levelt et al. 2006), and GOME-2 (Callies et al. 2000). All instruments fly in near-polar, sun-synchronous orbits,

---

[1] This process of NO formation is called *Zeldovich mechanism*. While for road transport, it accounts for 90–95 % of all emitted NO, this fraction depends strongly on the fuel type and is considerably lower for, e.g., combustion of coal.

Table 2.1 Key characteristics of the four satellite instruments used in this study

Instrument	Equator crossing	Global coverage (days)	Available period	Pixel (km^2)
GOME	10.30 a.m.	3	1995/10–2003/06	40 × 320
SCIAMACHY	10.00 a.m.	6	2002/08–now	30 × 60
OMI	1.30 p.m.	1	2004/10–now	13 × 24
GOME-2	9.30 a.m.	1.5	2007/01–now	40 × 80

leading to constant local overpass times for all measurements. The distinct characteristics of the four instruments are summarized in Table 2.1.

Combining different instruments' time series into one consistent trend estimate is challenging, as spatial resolution differences introduce inconsistencies. Previous studies therefore either artificially downgraded the higher-resolved measurements to yield comparable pixel sizes (Konovalov et al. 2010; van der A et al. 2008), or studied large areas, where instrumental differences tend to average out (Richter et al. 2005).

2.1.2 Method

The spectra of backscattered solar radiation as measured by the four instruments GOME, SCIAMACHY, OMI, and GOME-2 are analyzed to derive total slant column densities of NO$_2$ using the DOAS method (Platt and Stutz 2008). The stratospheric component of the measured signal is subtracted from the total columns using scaled NO$_2$ fields from the Bremen 3d Chemistry and Transport Model (CTM) (Sinnhuber et al. 2003b, c; Richter et al. 2005). The resulting tropospheric slant columns are then converted to vertical columns using radiative transfer calculations based on climatological data (Nüß 2005; Richter et al. 2005). Finally, a cloud filter based on the FRESCO+ cloud algorithm (Wang et al. 2008) is applied to select only those measurements with less than 20 % cloud cover.

The derived tropospheric vertical column densities (VCD$_{\text{trop}}$) of NO$_2$ are then binned globally to a 0.125° × 0.125° grid. For each instrument, daily and, subsequently, monthly averages are calculated. For each region of interest, one VCD$_{\text{trop}}$ NO$_2$, denoted by $Y_{(t,i)}$, is then calculated per time t (in years) and satellite instrument i by averaging over all grid cells within the region's boundaries.

To derive trend estimators for the temporal evolution of the linear and harmonic components of the measured VCD$_{\text{trop}}$, we assume that the measured values consist of a trend and a noise component (Mudelsee 2010):

$$Y(t, i) = X_{trend}(t, i) + S(i) \cdot X_{noise}(t, i) \tag{2.1}$$

Here, $S(i)$ denotes the standard deviation of all $Y_{(t,i)}$ for the given i (as a measure of the instrument's variability), and the trend component $X_{trend}(t, i)$ is defined by

$$X_{trend}(t, i) = \omega \cdot t + \mu_i + \eta_i \cdot (1 + \xi \cdot t)$$

$$\cdot \sum_{j=1}^{4} (\beta_{1,j} \cdot \sin(2\pi jt + \varphi) + \beta_{2,j} \cdot \cos(2\pi jt + \varphi)) \qquad (2.2)$$

ω is the linear annual growth rate, μ_i are the offsets of the linear trend per instrument, $\beta_{k,j}$ are the coefficients of the sine/cosine expansion to model the seasonal cycle, while η_i gives the amplitude of the seasonality, and ξ accounts for a trend in the amplitude of the seasonality.[2] For ease of notation, we define the parameter vector $\theta = (\omega, \mu_i, \eta_i, \xi, \beta_{k,j}, \varphi)$.

The corresponding trend estimator $\hat{\theta}$ is then calculated by minimizing the sum of the squared noise components using a modified Powell's method (Powell 1964).

2.1.3 Results

When the trend model (1) is fitted to data from large regions such as the central East coast of the USA, the agreement between the different instruments' measurements is expectedly good. This is reflected in the large similarity between the instrument-dependent trend parameters $\hat{\mu}_i$ and $\hat{\eta}_i$, as $\sigma(\hat{\mu}_i)/\bar{\hat{\mu}}_i = 0.03$ and $\sigma(\hat{\eta}_i)/\bar{\hat{\eta}}_i = 0.07$. The trend parameters found for this region are $\hat{\omega} = -4.7\,\%$ per year and $\hat{\xi} = -6.0\,\%$ per year.[3]

When applied to data from megacity regions, the differences between the instruments become more apparent. Generally, we find strong decreasing trends in the developed world, while developing megacities show strong increasing trends as expected. In virtually all cases the signs of $\hat{\omega}$ and $\hat{\xi}$ are the same. The magnitude of the harmonic trend estimator $\hat{\xi}$ however varies greatly. While for many regions, $\hat{\omega}$ and $\hat{\xi}$ are similar in magnitude (e.g. Beijing, Hong Kong, New York, Po Valley), very often the harmonic component $\hat{\xi}$ is significantly larger than $\hat{\omega}$, as e.g., in Athens, Baghdad, Barcelona, and Cairo. Of those regions considered in this study, only the BeNeLux region shows no clear change in the seasonality of the measured NO$_2$ signal.

Many regions do not show large differences between the instruments (e.g. Delhi), while in other regions, these differences are strongly pronounced, as in Tehran (see Fig. 2.1). The linear and harmonic components $\hat{\omega}$ and $\hat{\xi}$ for some selected megacity regions are summarized in Table 2.2.

---

[2] $\eta_1 \equiv 1$.

[3] In this study, all trend estimators are given relative to the average value over the whole time period 1996–2010.

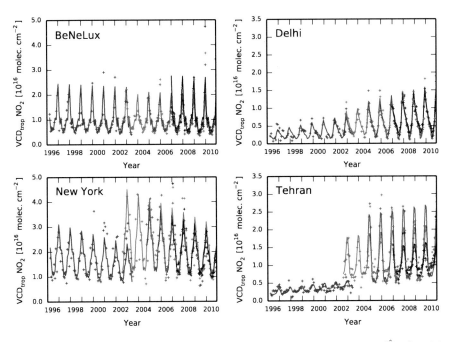

Fig. 2.1 Measured monthly averages $Y_{(t,i)}$ (*crosses*) and fitted trend function $X_{trend}(\hat{0},t,i)$ (*solid lines*) for the four instruments GOME (*blue*), SCIAMACHY (*red*), OMI (*green*), and GOME-2 (*black*)

Table 2.2 Anual growth rates of linear ($\hat{\omega}$) and harmonic ($\hat{\xi}$) trend components for selected megacity regions for the 1996–2010 period, relative to the period's average value

City	Linear trend ($\hat{\omega}$) (%/yr)	Harmonic trend ($\hat{\xi}$) (%/yr)
Athens	−2.6	−6.3
Baghdad	+9.2	+13.5
Barcelona	−2.3	−6.1
Beijing	+4.4	+5.9
BeNeLux	−2.2	−0.1
Cairo	+4.8	+13.0
Delhi	+5.0	+9.0
Hong Kong	+1.4	+1.1
New York City	−4.2	−5.6
Po Valley	−2.3	−1.5
Tehran	+4.4	+1.3
Tokyo	−3.5	−4.9

2.1.4 Discussion and Conclusion

The observed decreases in tropospheric NO$_2$ abundances throughout the developed world follow the improvements which have been made in reducing emissions and fighting air pollution. For example, NO_x emissions in Europe have reportedly dropped by 30 % from 1990 to 2000 and by 18 % from 1996 to 2002 (Lövblad et al. 2004). In the developing world, the increases in VCD_{trop} NO_2 reflect the strong growth in energy consumption, which goes along with rapid economic growth. China, for example, has seen a 64 % increase in NO_x emissions from 1996 to 2004 (Zhang et al. 2007).

The derived trend estimators show generally good agreement with previous trend studies (van der A et al. 2008; Konovalov et al. 2010). Perfect agreement however cannot be expected, because trend analyses of relatively short time series depend very strongly on the exact period covered and the used methodology.

While NO_2 shows a pronounced diurnal cycle,[4] the small timespan of e.g. 30 min between GOME and SCIAMACHY measurements cannot be the only reason for the observed differences between the four instruments. Model inter-comparison studies suggest that NO_2 levels at the instruments' overpass times differ by at most 10 % (Huijnen et al. 2010).

The spatial resolution of the satellite measurements however has significant effect on the observed NO_2 levels. The larger the pixel size, the more inhomo-geneous are the actual NO_2 levels in the troposphere within the area covered by one measurement, and the more the high pollution peaks of e.g. megacities are smeared out by the spatial integration of the instrument. Using topography (U.S. Geological Survey 2004) and population density (Center for International Earth Science Information Network et al. 2005) data, this can be seen very clearly using the example of the three cities Delhi, New York, and Tehran. In Delhi, which lies in a topographically flat region with homogeneously high population density, virtually no difference between the four instruments can be observed. Under these conditions, emitted NO_2 can spread without barriers. The area around New York City is also topographically flat, but the densely populated area (as a proxy for areas with high NO_x emissions) is restrained to the landmass, which makes up only 2/3 of the whole area. Therefore, emitted NO_2 can spread towards the ocean, leading to NO_2 column gradients within the observed area. This NO_2 gradient between megacity and open ocean leads to the noticeable impact of the instru-ment's pixel size on retrieved VCD_{trop}. In the case of Tehran, emissions are restrained to the city's boundaries, as it lies in a desert region. Moreover, the emitted NO_2 cannot spread evenly throughout the area, because Tehran is bordered by the Alborz mountain range towards the North and East, leading to inhomo-geneous NO_2 pollution in the observed area and thus to lower NO_2 columns in case of large pixel sizes.

We conclude that satellite observations are very well suited to observe tropo-spheric NO_2 pollution over megacities: decreasing and increasing trends can be observed in developed and developing countries, respectively. In spite of princi-pally identical measurement conditions all over the globe, special attention must be paid to the spatial resolution of the instrument and to the spatial patterns of

---

[4] In urban areas, the diurnal cycle of NO_2 is dominated by rush hour traffic peaks.

emissions and topography, as the selection of the measurement area, combined with differing spatial resolution of the available satellite instruments, strongly impacts the retrieved NO_2 values.

2.2 A Brief Example on the Application of Remotely Sensed Tracer Observations in Atmospheric Science—Studying the Impact of Stratosphere–Mesosphere Coupling on Polar Ozone Variability

Christoph G. Hoffmann[1] (✉), Matthias Palm[1], Justus Notholt[1], Uwe Raffalski[2] and Gerd Hochschild[3]

[1]Institute of Environmental Physics, University of Bremen, Germany
e-mail: christoff.hoffmann@iup.physics.uni-bremen.de
[2]Swedish Institute of Space Physics Kiruna, Sweden
[3]Institute for Meteorology and Climate Research, Karlsruhe Institute of Technology, Germany

Abstract This section outlines the realization and application of tracer observations in atmospheric science using an example in the context of ozone research; one particular factor contributing to polar ozone variability, namely the descent of mesospheric air into the stratosphere, is quantified. The estimation of this quantity, with results in a velocity of roughly 300 m/d, is outlined on the basis of ground-based microwave radiometry measurements of carbon monoxide in the Arctic.

Keywords Mesosphere–stratosphere coupling · Tracer observation · Ozone variability · Remote sensing · Microwave radiometry · Energetic particle precipitation · Atmospheric carbon monoxide

2.2.1 Introduction

The stratospheric ozone layer is one of the atmospheric key components of the climate system. It does not only absorb harmful ultraviolet (UV) radiation, but it also plays a major role for stratospheric temperatures and has indirect effects on the atmospheric circulation. The focus of ozone related research has been shifting from the understanding of the Antarctic ozone hole and Arctic ozone depletion towards the interaction of the (recovering) ozone layer and climate change in the last decade, demonstrating the importance of couplings between the individual layers of the atmosphere and other components of the climate system. On the one hand, climate change is expected to affect the ozone recovery (WMO 2007). On the other hand, the perturbed ozone layer influences the current climate, which has to be considered in climate change research. E.g., the global trend of increasing mean surface temperatures seems to be masked in the inner of the Antarctic continent by ozone hole related effects (Thompson and Solomon 2002).

To understand these interactions more accurately, it is necessary to understand both, the natural and the human-induced ozone variability in more detail. The refined knowledge is finally used to improve the performance of current climate models, which become more realistic by including more climate components and their mutual couplings.

Natural polar ozone variability is partly caused by the dynamical variability of the polar winter atmosphere, which affects ozone directly through transport or indirectly through changes of temperature and chemistry. The general aim of this study is the observation of the dynamical evolution of the polar winter stratosphere and mesosphere from the ground. This is conducted indirectly by measuring carbon monoxide (CO), which acts as tracer for dynamics in this region, using microwave remote sensing.

One of the processes particularly studied with this approach, is the energetic particle precipitation (EPP) indirect effect (Randall et al. 2009). It refers to the penetration of high energetic particles from the sun or from space into the mesosphere and lower thermosphere (MLT) region (at about 100 km altitude). There, they interact with atmospheric constituents and form NO_x among others. The meridional circulation in the mesosphere is directed towards the winter pole and leads to a descent of air above the winter pole, so that this NO_x can be transported into the polar winter stratosphere, where it contributes as catalyst to ozone depletion. The strength of the descent determines how much NO_x enters the stratosphere, and also how deep it descends, i.e., if it will reach the ozone layer. It is therefore shown in this study, how the strength of the descent can be estimated from CO observations.

2.2.2 CO as Tracer in the Polar Stratosphere and Mesosphere

CO is largely produced in the MLT region by the photolysis of CO_2. In the stratosphere and lower mesosphere, however, a loss reaction involving OH is dominant. The volume mixing ratio (vmr) profile of CO exhibits therefore a steep increase with altitude between both regions (Fig. 2.2). Furthermore, these chemical reactions are inactive in the absence of sunlight. Therefore CO has a long chemical lifetime particularly during polar night, so that changes of the vmr profile in these conditions must be due to dynamics. Thus, CO can there be used as tracer for dynamical processes (Solomon et al. 1985) and particularly for the descent of air above the winter pole, since the high CO values are shifted to lower altitudes by the winter descent (Fig. 2.2). In turn, from a time series of CO profiles, the properties of the vertical motion can be examined.

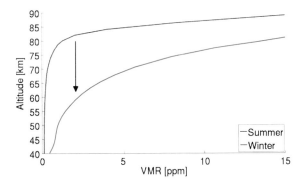

Fig. 2.2 Sketch of CO profiles during summer and winter demonstrating the downward shift of the higher CO vmr values as indicated with the *black arrow*

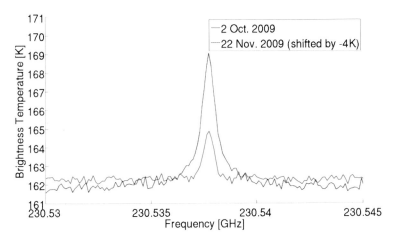

Fig. 2.3 Two examples of CO spectra. A relative change of the total CO abundance can be seen by eye (low CO abundance in *red* and high CO abundance in *blue*), whereas for the derivation of a CO profile a numerical process has to be performed, which separates the contributions from the single atmospheric layers

2.2.3 Measurements of CO by Microwave Radiometry

CO abundances of the stratosphere and mesosphere can be measured with several remote sensing techniques. Here we use ground-based microwave radiometry, which measures thermal emission of the atmosphere, thus can be used during polar night. Furthermore, this technique generally provides altitude-resolved information, so that vertical vmr profiles can be derived. The direct result of a single measurement is, however, not a vmr profile, but a microwave spectrum, showing the CO emission line at 230 GHz (Fig. 2.3). This line is a composite of the individual emissions of all atmospheric layers. The line shape of each individual emission is thereby dominantly controlled by the air pressure of the respective

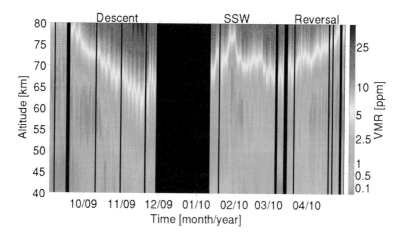

Fig. 2.4 Time series of KIMRA CO profiles showing all major features of dynamical variability throughout the winter 2009/2010

layer (pressure broadening), which is the key for the derivation of altitude-resolved data. The individual contributions to the composite have therefore to be deconvolved by a numerical process, the 'retrieval', to gain the vmr profile.

2.2.4 Interpretation of the KIMRA CO Time Series

The Kiruna Microwave Radiometer (KIMRA), used as example in this study, has been measuring microwave spectra of CO since 2007 in Kiruna, northern Sweden. The results of the retrieval of these spectra have recently been published (Hoffmann et al. 2011), showing that the measurement is sensitive between 40 and 80 km altitude. The corresponding time series of CO profiles for the winter 2009/2010 is shown in Fig. 2.4. The strong gradient of each CO profile is visible in the color-coding, and the descent of air during fall can be identified by the shift of similar colours from higher to lower altitudes. The average descent velocity can roughly be estimated by eye from the two altitudes of one particular CO level in October and in December, respectively, to be approx. 300 m/d (yellowish contours). This simple estimation works well for average velocities over a few months but is replaced by the analysis of the continuity equation of CO for more quantitative purposes.

Furthermore, the time series exhibits more CO variability starting at the end of January (Fig. 2.4), which corresponds to a dynamical perturbation known as sudden stratospheric warming (SSW). This strong dynamical perturbation is an important contribution to the variability of the polar winter stratosphere. Finally, the CO vmr decreases in spring, which is due to both, the reversal of the meridional circulation during the transit from winter to summer and the increasing exposure to sunlight in spring. All these dynamical features naturally influence the observed polar ozone variability, and are studied using such CO observations.

2.2.5 Summary and Conclusion

This section outlines the application of tracer observations in atmospheric science, showing how one particular factor can be quantified that finally contributes to the natural variability of the polar ozone layer. The initial interest is the strength of the descent of mesospheric air into the stratosphere and its representation in recent atmospheric models. Since the descent is not directly observable, a tracer for this motion, namely CO, is found based on its chemical properties. Since CO can also not be measured in situ, remote sensing is applied to measure spectra of the thermal emission of atmospheric CO from the ground. The CO vmr profiles are derived from these spectra, using a numerical retrieval technique.

The complete procedure has been outlined in this paper for the CO measurements of the radiometer KIMRA in the Arctic. The strength of the descent has been estimated to be roughly 300 m/d during the northern winter 2009/2010. A comparison of the presented CO time series with the Whole Atmosphere Community Climate Model (WACCM, Garcia et al. 2007) of the National Center for Atmospheric Research in Boulder, U.S.A. has been performed recently and has revealed a very good agreement (Hoffmann et al. 2012).

Acknowledgements The presented analysis of the CO data was supported by the German Research Foundation (Deutsche Forschungsgemeinschaft, DFG) with the projects NO 404/8-1 and PA 1714/3-2. The KIMRA instrument has initially been funded by the Knut and Alice Wallenberg Foundation. Substantial support for maintenance and development of the system was provided by the Swedish National Space Board and the Kempe Foundation.

2.3 Contamination of the Western Pacific Atmosphere

Theo Ridder[1] (✉), Justus Notholt[1], Thorsten Warneke[1] and Lin Zhang[2]

[1]Institute of Environmental Physics, University of Bremen, Germany
e-mail: jnotholt@iup.physik.uni-bremen.de
[2]School of Engineering and Applied Sciences, Harvard University, Cambridge, MA, USA

Abstract The Western Pacific is one of Earth's most remote areas and air in this region belongs to the cleanest air worldwide. However, the global anthropogenic and natural release of trace gases causes increasing contamination of Western Pacific air masses. To improve the knowledge about Western Pacific air, pollution carbon monoxide and ozone distributions were measured in the Western Pacific during a ship campaign with research vessel *Sonne* in autumn 2009. Between Japan and New Zealand observations of atmospheric trace gases were performed using solar absorption Fourier Transform infrared spectrometry. Distinct plumes of elevated carbon monoxide and ozone concentrations in the northern and southern hemisphere were observed. The cause of the measured air pollution is examined using a chemistry transport model showing that global scale transport of pollutants from Asia and Australia, but also partly from source regions as far away as Europe and Africa, causes the contamination of Western Pacific air. Generally, in the northern hemisphere a contamination of the Western Pacific air through fossil fuel combustion is dominant, while in the southern hemisphere trace gas emissions from biomass burning cause the most significant contamination.

Keywords   Western Pacific · Remote sensing · Atmospheric modelling · Chemical composition · Carbon monoxide · Troposphere · Airmass transport · Air pollution

2.3.1 Introduction

The Western Pacific is one of the most remote areas on Earth. Due to the remoteness air in this region belongs to the cleanest air masses worldwide. However, the global increasing anthropogenic emission of trace gases (Solomon et al. 2007) and the emission of trace gases from natural sources cause contamination of the atmosphere in the Western Pacific. This contamination can occur by a direct release of trace gases in this region through e.g. ship and flight traffic (Eyring et al. 2005). An even more important factor is the global transport of air pollutants from continental source regions into the Western Pacific. These continental source regions include e.g. industrial urban centers in South–East Asia (Jacob et al. 2003) and biomass burning regions in Africa (Andreae and Merlet 2001).

The Earth's system reacts delicately to contamination of the Western Pacific atmosphere. The tropical Western Pacific is considered to be the main region where tropospheric air masses are transported into the stratosphere (Holton et al. 1995). This upwelling process is associated with transport of trace constituents from the troposphere into the stratosphere. An enhanced entry of tropospheric trace gases into the Western Pacific, thus, follows an enhanced entry of pollutants into the stratosphere. This entry of tropospheric pollutants can disturb the chemical equilibrium of the stratosphere and can have a strong impact on stratospheric chemistry. Recent investigations deal with problems, such as whether the current entry of tropospheric pollutants into the stratosphere causes a delay in stratospheric ozone recovery (WMO 2007).

Measurements of atmospheric trace constituents in the Western Pacific are rare. Thus, the composition of the Western Pacific atmosphere is not very well known. Furthermore, the impact of individual sources and source regions on Western Pacific air quality is not fully understood. Here, we investigate Western Pacific air contamination and analyze the origin of the pollution.

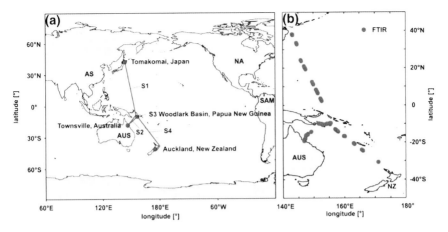

Fig. 2.5 **a** Overview of the ship campaign with RV *Sonne* in the Western Pacific divided into four cruise sections S1–S4. **b** Measuring points of the solar absorption FTIR spectrometer during the ship campaign

2.3.2 Method

To improve the knowledge about Western Pacific air contamination, atmospheric measurements of carbon monoxide (CO) and ozone (O_3) have been performed during a ship campaign with research vessel (RV) *Sonne* from Japan to New Zealand in autumn 2009. Aboard, solar absorption Fourier Transform infrared (FTIR) spectrometry (Rao and Weber 1992) was performed. The ship track is shown in Fig. 2.5. The measurements have been compared to a simulation with the full chemistry transport model *GEOS-Chem* for atmospheric composition (Bey et al. 2001) in order to evaluate the sources of the observed pollution.

2.3.3 Results

Figure 2.6 (top) shows the CO and O_3 distributions measured during the four sections of the ship campaign with RV *Sonne* (Fig. 2.5) in the Western Pacific. CO is presented as the total column amount, which describes the total amount of CO in the whole atmosphere above the observer. O_3 is pictured as the tropospheric column, which describes the total amount of O_3 in the troposphere above the observer. In comparison, the corresponding simulated CO and O_3 concentrations calculated with the *GEOS-Chem* model are presented showing that the simulations reproduce the observations well. Both distributions show several distinct plumes (PE1–PE3) of elevated CO and O_3 concentrations (PE1: beginning of cruise section S1, PE2: end of S1 and beginning of S2, and PE3: end of S3 beginning of S4).

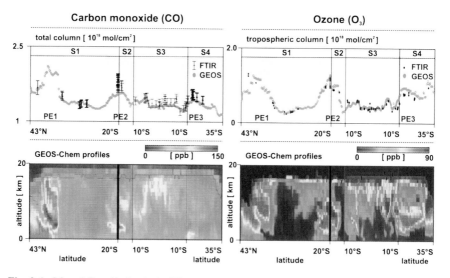

Fig. 2.6 CO and O_3 pollution in the Western Pacific measured during the ship campaign with RV *Sonne* and simulated with the *GEOS-Chem* model; *top* FTIR total and tropospheric column amounts compared to the *GEOS-Chem* model showing distinct plumes of elevated CO and O_3 (PE1–PE3), *bottom GEOS-Chem* vertical profiles showing the vertical distribution of the pollution

In order to get information about the vertical distribution of the contamination, the vertical profiles of CO and O_3 simulated with the *GEOS-Chem* model are shown in Fig. 2.6 (bottom). Pollution PE1 is distributed in the whole troposphere, pollution PE2 is apparent in the middle troposphere, and pollution PE3 occurs in the lower (CO) and middle (O_3) troposphere.

Within cruise section S3 elevated CO and O_3 in the upper troposphere is, furthermore, visible in Fig. 2.6 (bottom). Tropospheric air is vertically transported into the upper troposphere and can reach the stratosphere in this region. Thus, an example of the entry of tropospheric pollutants into the stratosphere is discovered.

2.3.4 Discussion and Conclusion

In order to determine the cause of the measured pollution in the Western Pacific, the *GEOS-Chem* model is used to calculate the contributions of various sources and source regions to the total pollution. Figure 2.7 shows the contribution of anthropogenic CO emissions from Asia and Europe (left) and contributions from biomass burning CO emissions (right) from Australia and Africa. Asian emissions are, according to this, the strongest CO source in the northern hemisphere with a direct impact on Western Pacific air quality. However, smaller European emissions are transported across Siberia towards the Western Pacific. North American anthropogenic emissions (not shown) only have a small impact on Western Pacific air composition.

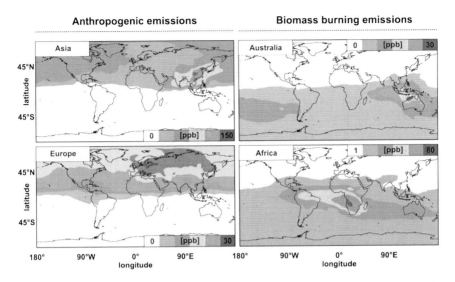

Fig. 2.7 Contributions from different sources and source regions to the measured pollution in the Western Pacific; *left* monthly mean (Oct. 2009) tropospheric average concentrations of anthropogenic CO emissions from Asia and Europe, *right* monthly mean (Oct. 2009) tropospheric average concentrations of biomass burning CO emissions from Australia and Africa

In the southern hemisphere the strongest source of CO pollution in the Western Pacific appears to be African biomass burning. Although most of African emissions are transported to the North-West, a significant plume of CO reaches the Western Pacific. In addition, biomass burning emissions in Australia considerably contribute to Western Pacific air contamination.

A detailed study about the contamination of the Western Pacific through CO and O_3 during the ship campaign with RV *Sonne* in autumn 2009 and an extensive evaluation of the sources and source regions of the measured pollution are published in Ridder et al. (2011).

2.4 Three Dimensional Model Simulations of the Impact of Solar Proton Events on Nitrogen Compounds and Ozone in the Middle Atmosphere

Nadine Wieters[1] (✉), Holger Winkler[1], Miriam Sinnhuber[2], Jan Maik Wissing[3] and Justus Notholt[1]

[1]Institute of Environmental Physics (IUP), University of Bremen, Germany
e-mail: nwieters@iup.physics.uni-bremen.de
[2]Karlsruhe Institute of Technology, Germany
[3]Department of Physics, University of Osnabrück, Germany

Abstract Solar Proton Events (SPEs) can have a large impact on the composition of the upper and middle atmosphere. These events are caused by fluxes of high energetic protons emitted from the sun that enter the Earth's atmosphere mainly in polar regions. These particle fluxes lead to ionization in the mesosphere and stratosphere. This ionization produces nitrogen NO_x (N, NO, NO_2) and hydrogen HO_x (H, OH, HO_2) compounds that can destroy ozone (O_3) very effectively. We investigate the impact of SPEs on stratospheric chemistry, in particular on NO_x and O_3, by means of model simulations with the Bremen 3-dimensional Chemistry and Transport Model (B3dCTM) for the period of the very large SPE in October/ November 2003. Our simulations indicate a large increase of NO_x of about 200 ppb during the event. Since NO_x is rather long-lived during polar winter, modeled values for the Northern Hemisphere are continually enhanced by 20– 100 ppb for several weeks after the event. These increased values are transported into the lower stratosphere within the polar vortex. During this downward transport, the enhanced NO_x values destroy up to 50 % of O_3 within the polar vortex, lasting until middle of December 2003. This O_3 depletion during the SPE 2003 can also be seen in observations.

Keywords Solar proton event · Energetic particle precipitation · Atmospheric modelling · Stratospheric chemistry · Ozone depletion · Polar vortex

2.4.1 Introduction

During Solar Proton Events (SPEs), huge solar eruptions, for instance, solar flares and Coronal Mass Ejections (CMEs), cause intense fluxes of protons with energies up to several hundred MeV. These particles can precipitate into the Earth's atmosphere along open magnetic field lines, therefore, mainly in polar regions. They can penetrate into different altitudes depending on their energies. Protons with energies of >1 MeV can enter the mesosphere, below 90 km, while protons with energies of >100 MeV can reach far into the stratosphere, below 40 km altitude. The precipitating particles lose their energy by collisions with high abundant atmospheric constituents such as nitrogen and oxygen, and produce ionization, dissociation, dissociative ionization, and excitation in the corresponding altitudes. The ionization and dissociation of N_2 leads to high amounts of NO_x (Porter et al. 1976; Rusch et al. 1981), whereas fast ion chemistry reactions produce high amounts of HO_x (Swider and Keneshea 1973; Solomon et al. 1981). These additional NO_x and HO_x compounds can destroy O_3 in catalytic reactions, whereby HO_x mainly destroys O_3 above 45 km and NO_x below 45 km (Lary 1997). Since energetic protons enter the atmosphere at both poles, these HO_x and NO_x productions occur in polar regions in both hemispheres. Because HO_x has a short lifetime (days), the impact on O_3 does not last very long, whereas the lifetime of NO_x constituents during polar night, in the absence of sunlight, is rather long (weeks). The produced NO_x can then be transported downward within the polar vortex which forms during polar winter. This means, since the air in the vortex is isolated from neighbouring latitudes, O_3 depletion along the downward transport of the NO_x is nearly undisturbed, and therefore, reaches far into the lower stratosphere. This makes SPEs during polar winter interesting to investigate, since the O_3 depletion in lower altitudes derives from a combination of solar activity and atmospheric dynamics. The impact of SPEs on NO_x and O_3 have been studied in models (e.g. Jackman et al. 2005; Sinnhuber et al. 2003b; Wissing et al. 2010; Funke et al. 2011), and observations (see e.g. Seppälä et al. 2004; López-Puertas et al. 2005). In this study we use a Chemistry and Transport Model (CTM) which is driven by meteorological data to investigate the very large Solar Proton Event on October/November 2003 (hereafter referred to as 'SPE 2003').

2.4.2 Model Simulations

To study the impact of SPEs on mainly NO_x and O_3, three-dimensional model studies have been performed with the B3dCTM. The model will be described briefly in the next section.

2.4.2.1 The B3dCTM

The Bremen 3-dimensional Chemistry and Transport Model (B3dCTM) has been developed as a combination of the chemistry transport model CTM-B (Sinnhuber et al. 2003a) with the chemistry code of the Bremen 2-dimensional model of the stratosphere and mesosphere (Sinnhuber et al. 2003b; Winkler et al. 2008). It computes the change in abundance of 58 species regarding chemical and dynamical forces. The dynamical forcing is prescribed by wind and temperature fields from, for instance, reanalysis of measurements, or modeled by general circulation models. The chemistry core computes about 180 neutral gas phase, photochemical, and heterogeneous reactions. The horizontal resolution is $3.75°$ in longitudinal and $2.5°$ in latitudinal direction. The vertical coordinate, in the model version used in this study, is potential temperature. This enables to calculate the vertical transport through diabetic heating and cooling rates. The vertical coverage is limited by the meteorological data set used. In this study, the ECMWF ERA-Interim (Dee et al. 2011) data set provided by the European Centre for Medium-Range Weather Forecasts (ECMWF) has been used. Hence, the vertical coverage of the model is 335–3231 K, about 10–60 km altitude.

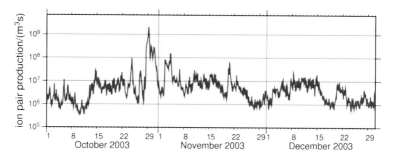

Fig. 2.8 AIMOS ionization rates due to proton precipitation averaged between 1 and 0.01 hPa over the Northern Hemisphere

2.4.2.2 Model Simulations of SPEs

To study the impact of SPEs, model simulations for the extremely large SPE on October/November 2003 have been performed. This SPE was one of the largest events in the past solar cycle. Huge CMEs have led to very high fluxes of energetic protons during October and November 2003, especially on 28/29 October and 3 November. To simulate the ionization impact of these particles, ionization rates derived from the Atmospheric Ionization Module Osnabrück (AIMOS) (Wissing and Kallenrode 2009) have been implemented into the B3dCTM. Figure 2.8 shows AIMOS ionization rates for this period averaged between 1 and 0.01 hPa over the Northern Hemisphere. The ionizing impact of the SPE is clearly seen in the mean ionization rates between 28 and 30 October 2003, and around the 4 November 2003, and is high during the whole period shown. Since the B3dCTM has a neutral chemistry core, the impact of the ionization needs to be parameterized. This was done by implementing production rates for NO_x and HO_x compounds based on Rusch et al. (1981), Porter et al. (1976) and Solomon et al. (1981). Accordingly, 1.25 NO_x (with 45 % N, and 55 % NO) and about 2 HO_x are produced per ion pair.

For this study, two model simulations have been performed. One model simulation assuming that no SPE occurred, hence no ionization rates included, and one with disturbed conditions, considering ionization rates based on precipitating protons.

2.4.3 Model Simulation Results

Figure 2.9 shows the impact of the SPE 2003 on HO_x (upper panel), NO_x (middle panel), and O_3 (lower panel). Each panel shows the difference of the two model runs described above. The values are differences of mean values within the band of 70–90° in high northern latitudes. The two upper panels of Fig. 2.9 show the direct impact of the ionization rates induced by the SPE as an increase of HO_x and NO_x,

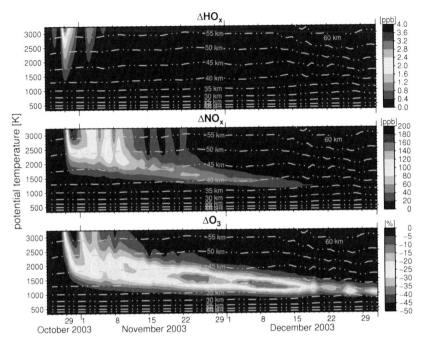

Fig. 2.9 Response in HO_x (*upper panel*), NO_x (*middle panel*), and O_3 (*lower panel*) to the SPE 2003 averaged over 70–90° North. See Sect. 3 for further details

largest at the days of highest ionization (see also Fig. 2.8). Increased HO_x values lead to O_3 depletion mainly above 45 km, which is also the main cause of the large modeled decrease of up to 50 % in O_3 during the first event around 30 October 2003. Since the SPE 2003 happened during northern polar winter, the directly produced NO_x was abundant for a long time, because of its long lifetime during polar night. During the event, the modeled NO_x is enhanced by about 200 ppb on 29/30 October 2003 and 100 ppb on 6 November 2003. Compared to background values assuming no SPE, this is an increase in NO_x by a factor of 500 and 200, respectively. A large fraction of the increased NO_x is then transported downwards within the polar vortex to lower altitudes, with enhancements of 40–80 ppb between 40 and 45 km lasting until the end of November 2003, and 20–60 ppb between 35 and 40 km lasting until middle of December 2003. The modeled NO_x enhancement between 35 and 45 km corresponds to an enhancement factor of 10–30, that leads to O_3 depletion of up to 40 %. Although the main O_3 depletion is diminished by the end of December, a small amount of NO_x values persists of up to several months after the event (not shown), leading then to an O_3 depletion of a few percent (see also Jackman et al. 2005).

2.4.4 Summary and Discussion

The impact of the SPE 2003 has been modeled with the B3dCTM. According to other model studies and observations (e.g. Seppälä et al. 2004; Jackman et al. 2005; López-Puertas et al. 2005; Wissing et al. 2010; Funke et al. 2011), the impact of this event is captured by the B3dCTM reasonably well. However, the impact on other species like HNO_3 (Verronen et al. 2008) and HCl (Winkler et al. 2009) are underestimated by a model that does not include any negative ion chemistry. This direct impact through negative ion chemistry needs to be parameterized additionally to the production rates described above. The influence of precipitating electrons on NO_x (Randall et al. 2007a, b; Sinnhuber et al. 2011) is not included here, but is currently under investigation. Model simulations (Wissing et al. 2010) indicate a direct impact of precipitating electrons on the ionization rates in the stratosphere, and therefore an impact on the production of NO_x in these altitudes. However, this issue is still not resolved since it has not been observed unambiguously yet (Funke et al. 2005; Clilverd et al. 2009; Funke et al. 2011). Model simulations, carried out with the B3dCTM with a similar setup, but including ionization rates due to proton and electron precipitation, seem to over-estimate the production of NO_x (Funke et al. 2011).

Acknowledgments This work is financially supported by the Deutsche Fors-chungsgemeinschaft (DFG) within the priority programme CAWSES (Climate and Weather of the Sun–Earth System). The ECMWF ERA-Interim data have been obtained from the ECMWF Data Server.

2.5 Evaluation of the Coupled and Extended SCIATRAN Version Including Radiation Processes Within the Water: Initial Results

Mirjam Blum[1,2] (✉), **Vladimir Rozanov**[2], **Tilman Dinter**[1,2,3], **John P. Burrows**[1] and **Astrid Bracher**[1,2,3] (✉)

[1]Institute of Environmental Physics, University of Bremen, Germany
e-mail: astrid.bracher@awi.de
[2]Helmholtz University Young Investigators Group PHYTOOPTICS
[3]Alfred Wegener Institute for Polar and Marine Research, Bremerhaven, Germany
e-mail: astrid.bracher@awi.de

To date, the software package SCIATRAN (Rozanov et al. 2002, 2005, 2008) has been used for modelling radiative processes in the atmosphere for the retrieval of trace gases from satellite data from the satellite sensor SCIAMACHY (Scanning Imaging Absorption Spectrometer for Atmospheric CHartographY onboard the satellite ENVISAT). This SCIATRAN version only accounted for radiative transfer within the atmosphere and reflection of light at the earth surface. However,

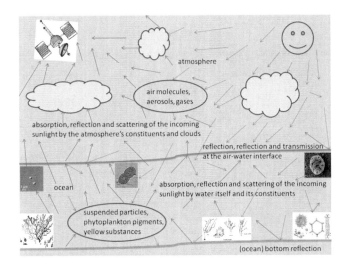

Fig. 2.10 Scheme of atmospheric and oceanic coupled radiative transfer

radiation also passes the air–water interface, proceeds within the water and is modified by the water itself and the water constituents. Therefore, SCIATRAN has been extended by oceanic radiative transfer and coupling it to the atmospheric radiative transfer model under the terms of established models for radiative transfer underwater (Kopelevich 1983; Morel et al. 1974, 2001; Shifrin 1988; Buitevald et al. 1994; Cox and Munk 1954a, b; Breon and Henriot 2006; Mobley 1994) and extending the data bases to include the specific properties of the water constituents (Pope and Fry 1994; Haltrin 2006; Prieur and Sathyendranath 1981) (Fig. 2.10).

So far, the coupling for the scalar radiative transfer is included. To analyse the quality of this new scalar coupled ocean–atmosphere radiative transfer version of SCIATRAN, model results of this and of the uncoupled SCIATRAN version are compared to satellite observations. In particular, we compared top of the atmosphere reflectances of the satellite sensor MERIS with the same parameters calculated with SCIATRAN by simulating the same optical conditions as for the satellite measurements. The MERIS data were selected in order to have different chl-a concentrations at different sites during different seasons. The main input parameters required for both SCIATRAN versions to simulate the selected MERIS collocations were concentrations of water vapour, ozone, chlorophyll, aerosol optical thickness, observation and illumination geometry, which were taken from the MERIS satellite and AERONET data sets. Both SCIATRAN versions consider the optical properties of organic and inorganic small (phytoplankton, bacteria, dust etc. <1 μm) and large (phytoplankton, zooplankton, sand etc. ≥1 μm) particles. Furthermore, in the coupled SCIATRAN version the single scattering albedo and the extinction coefficient can be set. Both properties of water particles have an

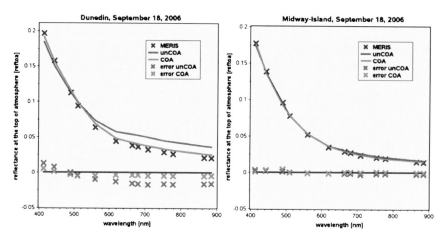

Fig. 2.11 Modelled (*red line* uncoupled, *green line* coupled SCIATRAN version) and measured (*blue cross* MERIS data) reflectances at the top of the atmosphere for (*left panel*) an oceanic region with chlorophyll concentration of 0.22 mg/m^3, near the aerosol station Dunedin (New Zealand, 163.47° E and 42.59° S), and (*right panel*) an oceanic region with chlorophyll concentration of 0.09 mg/m^3, near the aerosol station Midway Island (Atoll close to Hawaii, 177.35° W and 28.34° N), both on September 18, 2006. The *red crosses* show the deviation of the uncoupled version versus measured reflectance, and the green crosses the deviation of the coupled version versus measured reflectance in absolute values

important impact on the modelled result, but they are not often measured and not available from the MERIS satellite data at all. Therefore, common values based on theory and own tests have been used.

Figure 2.11 shows first results of these comparisons for two sites in the Pacific Ocean. Very good agreement (within 2–3 %) for both SCIATRAN versions was obtained with the satellite observation by MERIS measured close to Midway Island (Fig. 2.11 right). Deviations of the two SCIATRAN simulations are a bit larger (5–10 %) to the MERIS measurement close to the Dunedin station (Fig. 2.11 left). However, the deviations for both cases are at nearly all wavelengths smaller for the coupled SCIATRAN simulations, emphasizing the need to consider a coupled radiative transfer for retrievals of ocean color or atmospheric parameters over the ocean. Future studies will evaluate further the improved SCIATRAN version by focusing on MERIS collocations which overpass long-term-biooptical stations (e.g. BOSSOULE, MOBY) where more of the optical and geophysical parameters have been measured. Here the optical conditions under which MERIS observations were taken are clearer.

2.6 Improving the PhytoDOAS Method to Retrieve Coccolithophores Using Hyper-Spectral Satellite Data

Alireza Sadeghi[1] (✉), Tilman Dinter[1,2,3], Marco Vountas[1], Bettina Taylor[2,3] and Astrid Bracher[1,2,3]

[1]Institute of Environmental Physics, University of Bremen, Bremen, Germany
e-mail: sadeghi@iup.physics.uni-bremen.de
[2]Helmholtz University Young Investigators Group PHYTOOPTICS
[3]Alfred Wegener Institute for Polar and Marine Research, Bremerhaven, Germany

Abstract This study was dedicated to improve the PhytoDOAS method, which was established to distinguish major phytoplankton groups using hyper-spectral satellite data from SCIAMACHY. Instead of the usual approach of the PhytoDOAS single-target fit, a simultaneous fit of a certain set of phytoplankton functional types (PFTs) was implemented within a wider wavelength fit-window, called *multi-target fit*. The improved method was successfully tested through detecting reported blooms of coccolithophores, as well as by comparison of the globally retrieved coccolithophores with the global distribution of Particulate Inorganic Carbon (PIC). We analyzed eight years of SCIAMACHY data to investigate the temporal variations of coccolithophore blooms in a selected region within the North Atlantic, which is characterized by the frequent occurrence of intensive coccolithophore blooms. These data were compared to satellite total phytoplankton biomass, PIC concentration, sea-surface temperature, surface wind speed and modeled mixed-layer depth (MLD) in order to investigate the bloom dynamics based on variations in regional climate conditions. The results show that coccolithophore blooms follow the first total chlorophyll maximum and are in accordance with the PIC data. All three biological variables respond to the dynamics in wind speed, sea surface temperature and mixed layer depth. Overall, the results prove that PhytoDOAS is a valid method for retrieving coccolithophore biomass and for monitoring bloom developments in the global ocean.

Keywords Phyto-optics · SCIAMACHY · PhytoDOAS · Coccolithophores

2.6.1 Introduction

2.6.1.1 Motivation

The most suitable approach to monitor the global distribution of marine phytoplankton and to estimate their total biomass is by satellite remote sensing. Using ocean-color sensors, long-term records of aquatic parameters are provided on a global scale, with different applications (e.g., improving the understanding on ocean biogeochemistry and marine ecosystem dynamics; assessing fisheries productivity and the distribution of harmful algal blooms). Surveying the distributions and developments of marine phytoplankton on a global scale (e.g. Yoder et al. 1993) has been done by retrieving aquatic chlorophyll-a (*chl-a*), the common phytoplankton pigment, which is used as an indicator of phytoplankton biomass (Falkowski 1998). Several biooptical empirical algorithms (e.g., OC4v4 by O'Reilly et al. 1998) and semi-analytical algorithms (Carder et al. 2004) have been developed, relying on water-leaving radiance detected by satellite sensors, to retrieve the total *chl-a*. However, due to the phytoplankton biodiversity and differences in the optical properties and biogeochemical impacts of different phytoplankton groups, remote identification of phytoplankton functional types (PFTs; Nair et al. 2008) has been of great interest. These developed PFT-based retrieval methods (e.g. Sathyendranath et al. 2004) will improve the estimates of the total phytoplankton biomass, as well as deepen the understanding of oceanic biogeochemical cycles. However, all these attempts have been dependent on empirical methods and therefore on a large sets of in situ measurements. This has motivated the development of an alternative method for retrieving PFTs, called PhytoDOAS (Bracher et al. 2009). This method is essentially different from the other PFT ocean-color algorithms by retrieving the optical signatures of different PFTs within the backscattered spectra measured by the hyper-spectral satellite sensor SCIAMACHY (on-board ENVISAT: ENVIronmental SATellite of the European Space Agency, ESA). Testing the improvement of PhytoDOAS in retrieving more PFTs has been the main goal of this study.

2.6.1.2 The Importance of Coccolithophores

Sensitive responses of phytoplankton to environmental and ecosystem changes make them reliable indicators of the variations in climate factors. Coccolithophores are an abundant phytoplankton group with a wide range of impacts on the oceanic biogeochemical cycles and a significant influence on the optical features of surface water (Tyrrell et al. 1999). Coccolithophores emit dimethylsulfide (DMS, Andreae 1990), which is converted into the sulfur aerosols and cloud condensation nuclei (CCN), and thereby influence the climate and the Earth's energy budget (Charlson et al. 1987). Coccolithophore blooms are very important because they are frequently occurring (Holligan et al. 1993), and have unique biooptical and biogeochemical properties (Balch et al. 2004): coccolithophores are the main planktonic calcifiers in the ocean, characterized by building up calcium carbonate $CaCO_3$ plates, called *coccoliths* (Westbroek et al. 1985) and making a major contribution to the total content of suspended PIC in the open oceans (Milliman et al. 1993).

2.6.1.3 Objectives

The main interest of this study was to develop an appropriate method for quantitative remote sensing of coccolithophores using satellite data. This satellite-based method was used for monitoring temporal variations of coccolithophores on a global scale and by that for studying the impacts of variations of climate factors on marine phytoplankton. The study also aimed to identify the capacity and applicability of the method for climate research, i.e., phenological studies of phytoplankton dynamics to track the regional impacts of climate change (Winder 2010). In detail, coccolithophore dynamics in a selected region was monitored along with the variations of certain geophysical parameters, representing the regional climate. The retrieval method takes into account the phytoplankton absorption spectra, the existence of multiple PFTs and the pathlength of light penetrating in the water. These factors are often not considered in current biooptical methods based on band-ratio algorithms. Moreover, by retrieving only the differential absorption features, PhytoDOAS has the potential to obtain results on PFT *chl-a* also in coccolithophore rich region, where hyper-spectral variation are still visible. In this sense, retrieving coccolithophore blooms provides a reliable application to test the functionality of the improved PhytoDOAS method.

2.6.2 Material and Method

2.6.2.1 From DOAS to PhytoDOAS

PhytoDOAS was established to discriminate major phytoplankton groups based on their specific absorption footprints on the backscattered radiation from the ocean (Bracher et al. 2009). Differential Optical Absorption Spectroscopy (DOAS, Perner and Platt 1979), often used in atmospheric remote sensing, was applied to extract information on ocean optics; in particular on inelastic scattering (Vibrational Raman Scattering, VRS) of water molecules (Vountas et al. 2007) and on either diatoms or cyanobacteria specific absorptions. Within DOAS, from the ratio of earth shine backscattered radiation to extraterrestrial radiation, the imprints of differential absorption features of all relevant absorbers within the retrieved wavelength range are extracted. From the absorption cross-sections their differential parts are calculated by the *least square optimization*, which is used to fit the parameters attributed to the slant column densities of all relevant absorbers and the polynomial coefficients (accounting for elastic scattering and the low frequency variation in the absorption cross sections). The main outputs of PhytoDOAS are the fit factors for phytoplankton absorption and VRS, which are retrieved independently through two separate fits and are processed together to derive the *chl-a* concentration of the target PFT (Bracher et al. 2009).

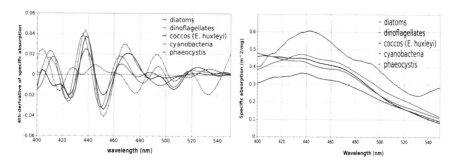

Fig. 2.12 Specific absorption spectra of five major PFTs (*right panel*) and corresponding fourth-derivative spectra (*left panel*) used for spectral analysis

2.6.2.2 Improvement to PhytoDOAS

As a recent improvement to PhytoDOAS, presented in this study, instead of just a single PFT target, three selected PFTs' absorption spectra (Fig. 2.12, left) are fitted simultaneously. These PFTs are diatoms, dinoflagellates and coccolithophores, which have been chosen through *derivative spectroscopy* analysis of phytoplankton absorption spectra (Aguirre et al. 2001). Based on the fourth derivative analysis, the appropriate PFTs (Fig. 2.12, right) for the simultaneous fit and also the appropriate fit-window for their retrieval (extended to 429–521 nm) have been selected. Accordingly, to incorporate the *multi-target fit* in PhytoDOAS, the single PFT absorption is replaced by several target PFTs' absorption spectra. The output includes a fit-factor assigned to each PFT target. The atmospheric reference spectra used in the PhytoDOAS *multi-target fit* are defined in (Bracher et al. 2009), the pseudo-absorption spectrum for VRS is defined in Vountas et al. (2007) and the target PFTs' spectra include the specific absorption spectra of diatoms, dinoflagellates and *E. huxleyi*, which were obtained by applying the PSICAM technique (Röttgers et al. 2007) on culture *E. huxleyi* and natural samples of diatoms and dinoflagellates. The *chl-a* conc. of PFTs existing in the natural samples were calculated by performing the CHEMTAX analysis (Mackey et al. 1996) on the high-performance liquid chromatography (HPLC) data.

2.6.2.3 Satellite and Modeled Data

For the PhytoDOAS method high spectrally resolved satellite data from SCIAMACHY have been used. This sensor covers a wide wavelength range (from 240 to 2380 nm) with a spectral resolution from 0.2 to 1.5 nm (Bovensmann et al. 1999). To compare and evaluate the PhytoDOAS coccolithophore results and to investigate their probable correlations with climate factors, four other satellite products were studied for a selected region in the North Atlantic from Jan. 2003 to Dec. 2010 as follows: (1) total *chl-a* from ESA's ocean-color dataset, GlobColour, providing merged data from three major ocean-color sensors: MODIS-Aqua, MERIS and SeaWiFS, (http://www.globcolour.info); (2) PIC data from MODIS-Aqua level-3 products (http://modis.gsfc.nasa.gov); (3) sea surface temperature (SST) from Advanced Very High Resolution Radiometer sensor (AVHRR: http://nsidc.org/data/avhrr); and finally (4) surface wind-speed data derived from the Advanced Microwave Scanning Radiometer-Earth Observing System (AMSR-E) sensor (http://remss.com). Additionally, a set of mixed layer depth (MLD) modeled data, obtained from the Ocean Productivity's merged FNMOC data (http://science.oregonstate.edu/ocean.productivity/index.php), were used as another regional geophysical parameter. The PFT assimilated data from the NASA Ocean Biochemical Model (NOBM, (Gregg and Casey 2007)), were used as a preliminary source of comparison of PETs.

2.6.3 Results and Discussion

2.6.3.1 Global Distribution of Coccolithophores

Figure 2.13 illustrates, as an example, the global comparison of PhytoDOAS coccolithophore *chl-a* to PIC conc. for Aug. 2005, showing consistent patterns between the two products. Since the concentration of PIC is proportional to the suspended *coccoliths* in surface waters, it is regarded as an indicator for coccolithophore*s* (Balch et al. 2005). In this sense, good agreements observed between the results of this study and PIC global distribution confirm the performance of the PhytoDOAS coccolithophore retrieval.

2.6.3.2 Time Series of Monitored Parameters in a Selected Region of the North Atlantic

To monitor the development of coccolithophore blooms, a region in the North Atlantic (named *nAtl*) at 53–63°N and 14–24°W of high bloom occurrence was selected based on the proposed global distribution of coccolithophores (e.g. Brown 1995) and on selected coccolithophore field studies (Holligan et al. 1993; Raitsos et al. 2006). Fig. 2.14 shows the monthly mean time series (from 2003 to 2010) of the PhytoDOAS coccolithophores and the other studied parameters: total *chl-a*, PIC, SST, wind-speed and modeled MLD. Table 2.3 shows the correlation coefficients over the bloom seasons in the *nAtl* region between the PhytoDOAS coccolithophores and the other parameters, and between PIC and other parameters.

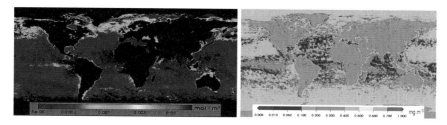

Fig. 2.13 PhytoDOAS coccolithophore *chl-a* from SCIAMACHY (*right panel*) and PIC conc. from the MODIS-Aqua (*left panel*) for Aug. 2005

2.6.4 Conclusions and Outlook

The PhytoDOAS results are consistent with the two other ocean color products, PIC conc. and total *chl-a*, and support the reported dependencies of coccolitho-phore biomass' dynamics to the compared geophysical variables. The results imply that coccolithophore blooms succeed the periods of elevated phytoplankton bio-mass, which are usually dominated by diatom blooms. Regarding the fact that the large reflectance from surface waters in coccolithophore rich areas affect the performance of the standard *chl-a* algorithms, the overestimations observed in the PhytoDOAS retrieved coccolithophores during the blooms (compared to the total *chl-a* product), suggest that over the coccolithophore bloom regions the ocean color sensors, supplying the GlobColour *chl-a* data, suffer from an underestimation of *chl-a*. However, the final assessment needs to be achieved by a future com-prehensive validation with in situ measurements. This study suggests that Phy-toDOAS is a valid method for retrieving coccolithophores' biomass and for monitoring bloom developments in the global ocean. Future applications of time-series studies using the PhytoDOAS data set are proposed, as using the new upcoming generations of hyper-spectral satellite sensors with improved spatial resolution. Regional adaptations of the PhytoDOAS coccolithophores are planned in order to account for the spatial variations in specific absorptions in respect to dominating coccolithophore species, e.g., following the well-known oceanic bio-geophysical provinces (Longhurst 1998).

Acknowledgements We are thankful to ESA, DLR, and the SCIAMACHY Quality Working Group (SQWG) for providing us with SCIAMACHY level-1 data. We thank NASA-GSFC for NOBM data and MODIS PIC images and data. We are grateful to NASA and ESA, particularly to the GlobColour project, for supplying satellite total *chl-a* data. We are thankful to AVHRR for providing the SST data, AMSR-E for the surface wind-speed products and also Ocean Produc-tivity for the monthly MLD data. Funding was provided by the HGF Innovative Network Funds (Phytooptics).

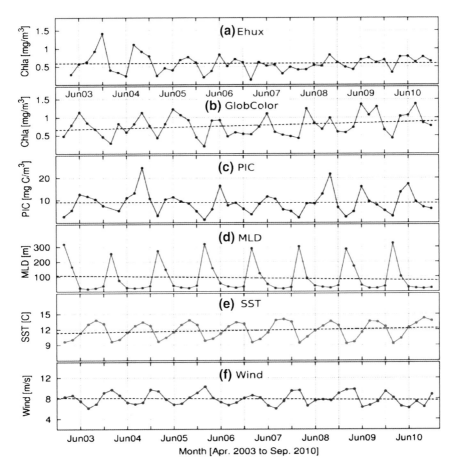

Fig. 2.14 Time series of six parameters monitored in North-Atlantic during the North Atlantic blooming season (January–September) from January 2003 to December 2010: **a** coccolithophores [Ehux] *chl-a* conc. retrieved by PhytoDOAS; **b** GlobColour total *chl-a*; **c** PIC conc. by MODIS-Aqua; **d** MLD from Ocean Productivity merged data; **e** SST from AVHRR; and **f** surface wind-speed from AMSR-E. The *dashed line* on each plot indicates the respective linear trend

Table 2.3 Correlation coefficients between the studied parameters in the North Atlantic region (*nAtl*)

	Ehux *chl-a* (PhytoDOAS)	Total *chl-a* (GlobColour)	PIC (MODIS -Aqua)	MLD	SST	Wind-speed
Ehux *chl-a*	–	0.53	0.66	−0.58	0.64	−0.49
PIC	0.66	0.79	–	−0.67	0.59	−0.72

2.7 Primary Productivity and Circulation Patterns Downstream of South Georgia: A Southern Ocean Example of the "Island Mass Effect"

Ines Borrione[1] (✉), **Olivier Aumont**[2] **and Reiner Schlitzer**[1]

[1]Alfred Wegener Institute for Polar and Marine Research, Bremerhaven, Germany
e-mail: ines.borrione@awi.de
[2]Laboratoire de Physique des Océans, Plouzané, France

Abstract Growth of phytoplankton in the Southern Ocean (SO) is largely limited by insufficient concentrations of the micronutrient iron, so that despite the large macronutrient reservoir, the SO is considered a High Nutrient Low Chlorophyll region. Therefore, phytoplankton growth is enhanced where exogenous iron is introduced to the system, for example downstream from islands. These confined regions sustain very rich ecosystems and are hot spots for atmospheric carbon dioxide drawdown. In this study, a combination of satellite derived measurements and model simulations are used to investigate the biological and physical environmental disturbances of the island of South Georgia (37°W, 54°S), which is located in the southwestern part of the Atlantic sector of the SO. We show not only that the island shelf is an important source of dissolved iron to the system, but also that the characteristic surface circulation patterns found downstream of the island play an important role in maintaining the shape and distribution of the developing phytoplankton bloom.

Keywords Southern Ocean · South Georgia · Island mass effect · Satellite observations · Primary productivity · High nutrient low chlorophyll regions · Biogeochemical modelling · Iron · ROMS · PISCES

2.7.1 Introduction

The Southern Ocean (SO, latitudes south of 40°S) covers 20 % of the global ocean, and surrounds the Antarctic Continent. The main hydrographic component is the Antarctic Circumpolar Current (ACC), an intense eastward flowing current encircling uninterrupted the continent.

The SO is a fundamental component of the Earth system and of its response to climate change (Marinov et al. 2006). Along portions of the Antarctic coast, the seasonal sea-ice formation generates intermediate and bottom waters which provide a major forcing to the global thermohaline overturning circulation, hence promoting heat, nutrient and gas fluxes with the other oceans.

Along the path of the ACC, upwelling of deeper waters replenishes surface waters with high concentrations of macronutrients (e.g., phosphates and nitrates) necessary for the growth of phytoplankton. However, due to insufficient concentrations of the trace metal iron, which is necessary for photosynthesis, algal growth in the SO is reduced, a reason why it is defined as a High Nutrient Low Chlorophyll (HNLC) region (Sarmiento and Gruber 2006).

Higher algal biomass is found where exogenous iron is introduced to the surface layers; among others, noteworthy sources are continental margins and atmospheric dust depositions (Tagliabue et al. 2009), as well as sea-ice melting (Lannuzel et al. 2007).

In situ investigations of the remote SO are generally limited to Austral summer, when most oceanographic cruises can be conducted. Consequently, satellite observations and modelling experiments are necessary tools to integrate with available in situ measurements. The former provide a quasi-synoptic view of regions as large as the SO, while the latter, albeit the necessary simplifications, allow for a virtual laboratory where to test hypothesis and better understand processes. Both tools were combined in the current study to investigate how and to what extent the sub-Antarctic Island of South Georgia generates an "Island Mass Effect", in other words, influences the surrounding physical and biogeochemical environment.

South Georgia (SG) is a relatively small island of the southwest Atlantic sector of the SO, lying between two of the major ACC fronts: the Polar Front (PF) to the north, and the Southern ACC Front (SACCF) to the south. Amidst HNLC waters, downstream from the island develops an intense phytoplankton bloom, which is clearly detectable from satellite ocean color imagery (Fig. 2.15, refer also to (Park et al. 2010)) and in situ measurements (Korb and Whitehouse 2004). This highly productive system sustains a rich ecosystem and one of the largest commercial krill fisheries (Atkinson 2001); furthermore, due to the enhanced phytoplankton growth, which promotes the biological carbon pump, it is identified as one of the most important Antarctic regions of atmospheric carbon drawdown (Schlitzer 2002).

2.7.2 Data and Methods

Chlorophyll-a concentration (chl-a) estimates were derived from the satellite Sea-viewing Wide Field-of-View Sensor (SeaWiFS) at 9 km resolution. Due to the large number of data gaps caused by frequent cloud cover, only monthly composites were used. Ocean color images were retrieved between November and February (austral summer) and between 1997 and 2010.

Surface circulation patterns were estimated from the AVISO Satellite Absolute Dynamic Topography (ADT) measurements. Weekly products were retrieved, and averaged to form a monthly climatology corresponding to the same time period of the SeaWiFS observations.

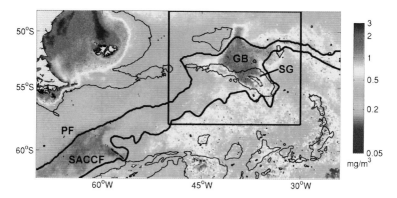

Fig. 2.15 November–February chlorophyll-a climatology from 1997 until 2010 in the southwest Atlantic sector of the SO from satellite ocean color estimates. *Bold lines* indicate the Polar Front (PF) and the Southern ACC Front (SACCF), *SG* indicates the island of South Georgia, and *GB* stands for Georgia Basin, the *rectangle* depicts the South Georgia region; *thin black lines* indicate the 2000 m isobaths

Model simulations were carried out with a coupled configuration of ROMS, a free surface, topography following, primitive equation regional model (Shchepetkin and McWilliams 2005) and PISCES, a 24-compartment biogeochemical model, where the cycles of carbon and the main nutrients (including iron) are resolved (Aumont and Bopp 2006). Simulations were run at 1/6° resolution ($\sim$ 12 km).

2.7.3 Results and Discussion

According to the climatology shown in Fig. 2.15, a large phytoplankton bloom develops in the South Georgia region; the position of the bloom appears to be confined over the Georgia Basin while outside the basin borders, chl-a remains mostly below 0.5 mg/m^3 hence indicating typical HNLC conditions.

Measurements of ADT in the SG region, allow for estimating the main pathways and intensities of surface circulation; the former are parallel to ADT contour lines, while the latter are more intense when contour lines are closer. As indicated by ADT contours (black lines in Fig. 2.16a), the flow encounters the island from the southwest and then continues northeast embracing the Georgia Basin at all sides. Furthermore, it is possible to infer intense currents along the borders of the basin (especially along the northern periphery, where contour lines are closely spaced) and a weaker circulation regime in the region found directly above the basin.

The similarity between surface circulation patterns depicted in Fig. 2.16a and the bloom distribution shown in Fig. 2.15 suggests that the surface circulation in the SG region plays an important role in maintaining the shape and position of the

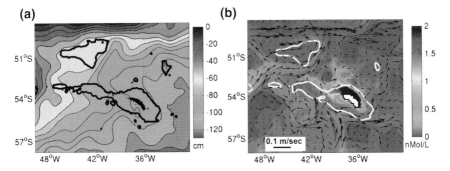

Fig. 2.16 **a** Absolute dynamic topography in the South Georgia region between November and February according to the 1997–2010 climatology. *Contour lines* indicate direction of main flow and distance between isolines provides an estimate of flow intensity. **b** Simulated distribution of surface dissolved iron (*color code*) and circulation (*arrows*) during December of model year 3. In both panels, 2000 m isobaths are indicated with *bold contour lines*

phytoplankton bloom developing downstream from SG. Therefore, circulation patterns are such that phytoplankton cells remain entrained in a favorable region where the flow is weaker and nutrient reservoirs may be continuously replenished.

The importance of SG as a source of dissolved iron is suggested by model simulations (Fig. 2.16b), which show the presence of a plume originating from the island that follows the direction of the main flow (arrows). However, the characteristic circulation patterns depicted in Fig. 2.16a remain difficult to resolve, possibly due to the sensitivity of the system to the adopted boundary conditions or the chosen model resolution which does not allow for a full representation of the physical environment (i.e., eddies, which at the chosen resolution are not resolved).

2.7.4 Conclusions

Although our results indicate that downstream from SG the observed phytoplankton bloom results from a characteristic physical environment, being SG a source of dissolved iron, a comprehensive understanding of the system, where a patchwork of top-down (i.e., grazing) or bottom-up (light or nutrient co-limitation) controls come into play, requires further in situ investigations and model experiments. The former would provide additional biogeochemical measurements which to date remain scarce, while the latter would help assess the relative importance of physical processes other than surface circulation (for example performing simulations where tidal effects and/or eddies are included) or additional nutrient sources in determining the observed biological response to the South Georgia "Island Mass Effect".

2.8 Summer Sea Ice Concentration Changes in the Weddell Sea and Their Causes

Sandra Schwegmann[1] (✉), **Ralph Timmermann**[1], **Rüdiger Gerdes**[1,2] **and Peter Lemke**[1,3]

[1]Alfred Wegener Institute for Polar and Marine Research, Germany
e-mail: sandra.schwegmann@awi.de
[2]Jacobs University, Bremen, Germany
[3]University of Bremen, Germany

Abstract Sea ice concentration, the fraction of sea ice coverage per unit area, is subject to regional climate variability in the Weddell Sea, Antarctica. The magnitude and origin of local trends in ice coverage, which are found to be most pronounced in summer, were analyzed using the "bootstrap algorithm sea ice concentration data" from the NSIDC for 1979–2006. The impact of atmospheric forcing such as air temperature and wind speed as well as that of ice drift obtained from satellite data and from free drift model simulations was studied. Generally, most of the observed sea ice concentration changes in summer can be related to wind changes. Also the temperatures correlate well with ice concentration, but whether sea ice concentration forced the temperature changes or vice versa is still unclear.

Keywords Sea ice · Concentration · Ice drift · Weddell Sea · Antarctic, Southern Ocean · SSM/I · Polar Pathfinder sea ice motion · Free drift · NCEP

2.8.1 Introduction

Sea ice is an indicator for climate change. In the Northern Hemisphere, ice extent (IE) has decreased over the last decades and the causes for this decrease have been subject to several studies (e.g. Parkinson and Cavalieri 2002; Stroeve et al. 2005). However, in the Southern Ocean, IE has increased by about 1 % per decade (Cavalieri and Parkinson 2008). This trend is a circumpolar average over different signals from various Antarctic regions. Here, we focus on the Weddell Sea sector.

Cavalieri and Parkinson (2008) found that the IE has increased during summer and fall by 6.3 and 1.8 %, respectively, and decreased during winter and spring by 0.1 %, but they were not able to fully identify the underlying mechanisms. We present an analysis of the sea ice concentrations (SIC) in the Weddell Sea for 1979–2006, showing that IE changes were a product of different trends in different parts of the Weddell Sea, namely (1) increasing SIC (and therefore a higher extent) in the eastern regions and (2) decreasing SIC along the Antarctic Peninsula. Our analysis indicates that wind and related drift changes caused SIC trends and that their influences differ between the marginal and coastal ice zones and the pack ice in the central Weddell Sea.

2.8.2 Data

SIC data with a resolution of 25 km, projected onto a polar stereographic grid, are provided by the National Snow and Ice Data Center (NSIDC) as a combination of observations from the Scanning Multichannel Microwave Radiometer (SMMR) and the Special Sensor Microwave/Imager (SSM/I) generated using the bootstrap algorithm (Comiso 1999). Data were correlated to the NCEP/NCAR reanalysis sea level pressure, 2 m air temperature, and 10 m wind from 1979 to 2006 and to sea ice drift data from the Polar Pathfinder Sea Ice Motion vectors (Fowler 2003, obtained from the NSIDC), for 1988–2006. Uncertainties of this product for the Weddell Sea were quantified through a comparison with buoy data by Schwegmann et al. (2011) and have been shown to increase in summer. Therefore, we also used velocities from a free drift model, forced with 10-wind speeds and ocean near-surface currents from the Finite Element Sea ice-Ocean Model (FESOM, Timmermann et al. 2009).

2.8.3 Results and Discussion

Mean SIC (Fig. 2.17) and their trends (Fig. 2.18) were analysed in this study. The strongest changes of SIC occurred in austral summer, with an increase at the eastern boundaries of the ice cover and a strong decrease along the coast of the Antarctic Peninsula (AAP, Fig. 2.18). The changes at the AAP are observed year-round and can be related to increased wind speeds in most months. For all months, not only in summer, the wind velocities are anti-correlated with SIC in this region (Fig. 2.19). The zonal wind is thereby anti-correlated with SIC during all three months, whereas the correlation between the meridional component and SIC alternates in sign (not shown).

Generally, the same anti-correlation as for wind is found between observed SIC and modelled free drift next to the tip of the AAP (not shown). Mean sea ice drift at this location is offshore, transporting ice from the coast into the central Weddell Sea and the marginal sea ice zones. Thus, if sea ice drift is enhanced due to increased wind speeds, the near coastal region loses ice, while at the same time ice coverage is increased in the marginal sea ice zones. In the inflow area in the eastern Weddell Sea, free drift speeds are mainly anti-correlated to SIC. A feasible explanation is that with stronger drift, more sea ice can be transported into the Weddell Sea basin and less ice remains in the north-eastern coastal region, while refreezing only gradually increases the ice coverage in this region.

Fig. 2.17 Mean sea ice concentrations for January, February and March for the period 1979 through 2006 in %

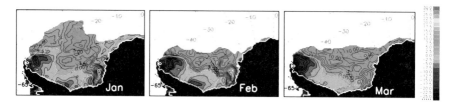

Fig. 2.18 Sea ice concentration trends in % concentration per decade

Fig. 2.19 Correlation between sea ice concentrations and wind speeds for January, February and March 1979–2006. *Vectors* show the mean wind field

Fig. 2.20 Temperature trends for January, February and March 1979–2006

Next to the wind and the free drift, also temperatures show a significant signal: In February and March, 2 m air temperatures are mainly anti-correlated to SIC, with coefficients of up to −0.8 in the marginal sea ice zone (not shown). Specifically the decreasing SIC at the AAP comes along with increasing temperatures. Trends of the 2 m air temperatures (Fig. 2.20) at the AAP are indeed positive and are certainly connected to the observed decrease in SIC. In addition, large parts of the Weddell Sea show a decrease in 2 m air temperatures in austral summer, which is consistent with increasing SIC in the marginal sea ice zone.

However, given that true observations in this region are sparse and that near-surface temperatures in the reanalysis datasets are bound to be affected by the SIC prescribed as a boundary condition in the assimilations system, it is impossible to tell whether warmer air forces ice to melt, or reduced ice coverage causes an increased ocean-to-air heat flux and thus a local warming of the lower atmosphere. An analysis of correlations with time lag of ± 1 month does not help to answer this question. In January, correlation of temperature changes one month behind those of SIC reveals a high anti-correlation (which indicates that reduced ice coverage causes a warming of the lower atmosphere), but for the other months, the highest anti-correlations occur without any time lag.

In addition, the analysis of cloud coverage does not yield a robust correlation to SIC, which may indicate that downward radiation fluxes do not play a major role in regional climate change in the Weddell Sea. Satellite-based ice drift patterns in summer are based on very few data points and show only little connection to the observed SIC changes.

2.8.4 Conclusion

Analyzed trends in near-surface wind and the closely connected ice drift are consistent with the observed decrease of SIC at the tip of the AAP together with the increased ice coverage in the eastern marginal sea ice zone. Correlation coefficients, however, are not very high, so that other effects certainly contribute to the observed changes. The finding that the correlations between simulated free drift and observed SIC are sometimes smaller than those for the reanalysis wind may result from the prescribed ocean currents as well as from the fixed mean sea ice thickness in the free drift model (although neither varying the prescribed mean sea ice thickness between 0.5 and 2 m nor using time-invariant ocean currents notably change the mean drift speeds), but it may also indicate that the advection of warm air with increased westerly winds adds to the purely dynamic effect. The weak connection between satellite-based ice drift and observed SIC may be due to uncertainties in the used satellite product, but further investigations on this topic are necessary.

2.9 Validation of the Snow Grain Size Retrieval SGSP Using Six Ground Truth Data Sets

Heidrun Wiebe[1] (✉), Georg Heygster[1] (✉) and Eleonora Zege[2]

[1]Institute of Environmental Physics, University of Bremen, Germany
e-mail: heygster@uni-bremen.de
[2]B.I. Stepanov Institute of Physics of the National Academy of Sciences of Belarus
(IP- NASB)Minsk, Belarus

Abstract For climate modelling, the snow grain size is an important parameter to determine the snow albedo on ground, which in turn affects the radiative balance of the Earth. Recently, the Snow Grain Size and Pollution (SGSP) retrieval from reflectances of the optical satellite sensor Moderate Imaging Spectroradiometer (MODIS), applicable for polar regions, was developed and implemented. In this section, a comparison of the SGSP-retrieved with ground-measured snow grain size is presented, using six different ground truth data sets from the Arctic, the Antarctic, Greenland, and Japan. It shows a good agreement with a correlation coefficient of 0.86 and with small differences below 15 % for almost half of the comparison cases.

Keywords Snow grain size · Retrieval · MODIS · Validation

2.9.1 Introduction

Snow is part of the cryosphere in the climate system of the Earth. It covers up to 45.000 km^2 of the Earth's surface. As in most cases the snow falls on a darker surface, a much larger amount of the incoming solar radiation is reflected, affecting the radiative balance of the Earth.

From the microphysical perspective, a single snow grain is an ice crystal consisting of frozen water and is formed in clouds. The single crystals cluster together and fall as snow flakes onto the Earth's surface. There, they accumulate to a layer of snow consisting of ice, air, and sometimes impurities like dust or soot. Inside the snow layer, metamorphism changes the shape and size of the ice crystals, mainly controlled by the temperature and the vertical temperature gradient.

Light falling on a snow layer mostly gets scattered, reflected, and partly absorbed. As snow is a porous medium of ice, air, and possibly impurities, the light entering a snow layer is often multiply scattered. The albedo of snow (as the ratio of reflected (including multiply scattered) and incident light) is high in the visible (1.0–0.9) and decreases towards the near-infrared (0.9–0.2). For a semi-infinite snow layer, the albedo is mainly determined by the size and shape of the grains, and the amount of impurities. Therefore, those parameters are important if modelling the Earth's climate.

Regular observations of the snow on a global scale can only be achieved by means of satellite remote sensing, as many snow covered areas in polar regions are difficult to access. The type of electromagnetic radiation sensitive to the snow grain size and impurities is the optical (visible and near-infrared) radiation in the range from 0.4 to 1.4 µm. Data in this spectral range are available from various satellite sensors at a spatial resolution in the order of 10 m–1 km.

An algorithm that uses optical observations to determine the snow grain size and impurity amount (here: soot concentration) with satellites is the Snow Grain Size and Pollution (SGSP) algorithm developed by Zege et al. (1998, 2008, 2011).

In this section, a validation study of the SGSP-retrieved grain size is summarized, using six ground truth data sets from different regions (Arctic, Antarctic, Greenland, and Japan) on different subsurfaces (land, sea ice, and lake ice) in the years 2001–2009. The full details are given in Wiebe et al. 2011.

2.9.2 The SGSP Retrieval

The SGSP algorithm, developed by Zege et al. (1998, 2008, 2011) computes the snow grain size and impurity amount from optical satellite data (here: MODIS Channels 3, 2, and 5 at 0.47, 0.86, and 1.24 µm). It uses a snow reflectance model based on an analytical asymptotic solution of the radiative transfer theory and on geometrical optics for the optical properties of snow (Zege et al. 2008):

$$R_i(\theta, \theta_0, \varphi) = R_0(\theta, \theta_0, \varphi) \exp\left(-A\sqrt{\gamma_i a_{ef}} \frac{K(\theta)K(\theta_0)}{R_0(\theta, \theta_0, \varphi)}\right) \tag{2.3}$$

$$\text{with } \gamma_i = 4\pi(\chi_i + \kappa C_S)/\lambda_i \text{ and } K(\theta) = 3\left(1 + 2\cos\theta\right)/7$$

- i refers to the MODIS Channel number i
- R_i is the snow reflectance received by the instrument
- R_0 is the bidirectional reflectance distribution function (BRDF) for non-absorbing snow
- θ, θ_0, and φ are the sensor zenith, solar zenith, and relative azimuth angle
- A is a form factor depending on the snow crystal shape (here: $A = 5.8$)
- γ_i is the absorption coefficient of snow
- a_{ef} the effective optical snow grain size
- K is the escape function
- χ_{-i} is the imaginary part of the complex refractive index of ice
- κ depends on the type of pollutant (here for soot: $\kappa = 0.2$)
- C_S is the relative volumetric soot concentration
- λ_i s the wavelength

Equation (2.3) is valid for retrievals directly on the surface, or for atmospherically corrected reflectances from satellite sensors. The atmospheric correction used in this work is described in Zege et al. (2011). Furthermore, two steps of data pre-processing are applied to the MODIS input data: removing striping artifacts due to instrument calibration errors and extracting snow pixels from non-snow pixels like water, soil, or clouds (Wiebe 2011).

The SGSP retrieval has two main characteristics distinguishing it from existing snow grain size retrievals (e.g. Nolin and Dozier 2000; Stamnes et al. 2007):

- Its reduced dependency on the snow crystal shape allows better results as the crystals may have different shapes (columns, plates, fractals) of different reflection characteristics, where the shape information is not available a priori on large areas in polar regions.
- Its validity at high solar zenith angles up to 75° (i.e. sun elevation down to 15°) allows applying the retrieval in polar regions where the sun is often low.

2.9.3 Validation Studies

The SGSP-retrieved snow grain size has been validated with ground truth data from six different measurements campaigns.

Aoki et al. (2007) measured the microphysical snow grain size by a handheld lens and ruler at several places on Hokkaido, Japan, and at Barrow, Alaska, in the years from 2001 to 2005. Gallet et al. (2010) measured the specific surface area (can directly be related to the optical snow grain size) at Dome C and on the traverse from Dome C to Dumont D'Urville, Antarctica, in the austral summer 2008/2009. Markus et al. (2006, Personal communication) measured the microphysical snow grain size on the sea ice of the Chuckchi Sea, Alaska. Scambos et al. 2007 measured the optical snow grain size by a ground spectrometer on the sea ice near the coast of Wilkesland, Antarctica. Painter et al. (2007) measured the optical snow grain size by a ground spectrometer at Swiss Camp near the west coast of Greenland. Brandt et al. (2008) measured the spectral albedo (can directly be related to the optical snow grain size) by a ground spectrometer, on the Elson Lagoon, Alaska.

The results of the six comparisons are shown in the scatter plot of Fig. 2.21, having a correlation coefficient R = 0.86. There are 17 cases with small differences below 15 %, 15 cases with intermediate differences between 15 and 50 %, and 5 cases with large differences above 50 %.

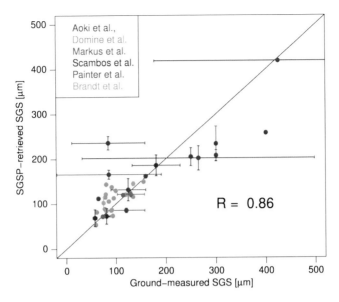

Fig. 2.21 Scatter plot of the SGSP-retrieved versus the ground-measured snow grain size for the 6 different data sets. *Dots* and *error bars* are local daily averages and their standard deviation. For cases without error bars only one measurement was available or the error was too little

2.9.4 Discussion and Conclusion

The comparison of the SGSP-retrieved to the ground-measured snow grain size from the six ground truth data sets shows a good agreement. Most cases of larger differences are influenced by cirrus clouds, wet snow, surface hoar, or wind crust. Discrepancies may also be attributed to SGSP retrieval, e.g. by instrument noise in the MODIS input data. Furthermore, the ground measurements are point measurements, whereas one satellite pixel has a size of 500×500 m^2 inside of which the snow may not be distributed homogeneously.

In the Aoki et al. (2007) data set, measuring the microphysical snow grain size, the ground observations have a large spread at one site and day (large error bars in Fig. 2.21) and larger discrepancies to the SGSP-retrieved grain size. Therefore, ground measurements of the optical snow grain size are preferable for validating satellite retrievals.

The final result of this work is that the SGSP retrieval has been validated successfully with a versatile mixture of ground measurements. It is implemented in a near-real time processing chain allowing efficient processing of the snow grain size, which influences the albedo and thus the radiative balance of the Earth.

References

Aguirre-Gomez R, Weeks AR, Boxall SR (2001) The identification of phytoplankton pigments from absorption spectra. Int J Remote Sens 22(2):315–338

Andreae MO (1990) Ocean–atmosphere interactions in the global biogeochemical sulfur cycle. Marine Chem 30:1–29

Andreae MO, Merlet P (2001) Emission of trace gases and aerosols from biomass burning. Global Biogeochem Cycles 15:955–966

Aoki T, Hori M, Motoyoshi H, Tanikawa T, Hachikubo A, Sugiura K, Yasunari T, Storvold R, Eide H, Stamnes K, Li W, Nieke J, Nakajima Y, Takahashi F (2007) ADEOS-II/GLI snow/ice products—part II: validation results using GLI and MODIS data. Remote Sens Environ 111:274–290. doi:10.1016/j.rse.2007.02.035

Atkinson A, Whitehouse MJ, Priddle J, Cripps GC, Ward P, Brandon MA (2001) South Georgia, Antarctica: a productive, cold water, pelagic ecosystem. Mar Ecol Prog Ser 216. doi:10.3354/meps216279

Aumont O, Bopp L (2006) Globalizing results from ocean in situ iron fertilization studies. Glob Biogeochem Cycles 20. doi:10.1029/2005GB002591

Balch WM (2004) Re-evaluation of the physiological ecology of coccolithophores. In: Thierstein HR, Young JR (eds) Coccolithophores. From molecular processes to global impact. Springer, Berlin, pp 165–190

Balch WM, Gordon HR, Bowler BC, Drapeau DT, Booth ES (2005) Calcium carbonate measurements in the surface global ocean based on moderate-resolution imaging spectrora-diometer data. J Geophys Res 110:C07001

Bey I, Jacob DJ, Yantosca RM, Logan JA, Field BD, Fiore AM, Li Q, Liu HY, Mickley LJ, Schultz MG (2001) Global modeling of tropospheric chemistry with assimilated meteorology: Model description and evaluation. J Geophys Res 106:23073–23095

Bovensmann H, Burrows JP, Buchwitz M, Frerick F, Noël S, Rozanov VV (1999) SCIAMACHY: mission objectives and measurement modes. J Atm Sci 56:127–150

Bracher A, Vountas M, Dinter T, Burrows JP, Röttgers R, Peeken I (2009) Quantitative observation of cyanobacteria and diatoms from space using PhytoDOAS on SCIAMACHY data. Biogeosciences 6:751–764

Brandt R, Gerland S, Pedersen C, Berntsen T, Borgar A (2008) Spectral albedo of the snow surface: Elson Lagoon, Barrow, AK. http://www.atmos.washington.edu/sootinsnow/PDF_Documents/Brandt_et_al_2005.pdf. Accessed 18 Oct 2010

Breon FM, Henriot N (2006) Spaceborne observations of sun glint reflectance and modeling of wave slope distributions. J Geophys Res 111:C06005. doi:10.129/2005JC003343

Brown CW (1995) Global distribution of coccolithophore blooms. Oceanography 8(2):59–60

Buitevald H, Hakvoort JHM, Donze M (1994) The optical properties of pure water. SPIE Ocean Optics XII 2258:174–183

Burrows JP, Weber M, Buchwitz M, Rozanov VV, Ladstaetter-Weissenmayer A, Richter A, DeBeek R, Hoogen R, Bramstedt K, Eichmann K-U, Eisinger M (1999) The global ozone monitoring experiment (GOME): mission concept and first scientific results. J Atm Sci 56:151–175

Callies J, Corpaccioli E, Eisinger M, Hahne A, Lefebvre A (2000) GOME-2—Metop's second-generation sensor for operational ozone monitoring. ESA Bulletin 102:28–36

Carder KL, Chen FR, Cannizzaro JW, Campbell JW, Mitchell BG (2004) Performance of the MODIS semi-analytical ocean color algorithm for chlorophyll-a. Adv Space Res 33:1152–1159

Cavalieri DJ, Parkinson CL (2008) Antarctic sea ice variability and trends, 1979–2006. J Geophys Res (Oceans) 113:C07004. doi:10.1029/2007JC004564

Center for International Earth Science Information Network (CIESIN) (2005) Columbia University, United Nations Food and Agriculture Programme (FAO), and Centro International de Agricultura Tropical (CIAT). Gridded Population of the World: Future Estimates

(GPWFE). Palisades, NY: Socioeconomic Data and Applications Center (SEDAC). Columbia University

Charlson RJ, Lovelock JE, Andreae MO, Warren SG (1987) Oceanic phytoplankton, atmospheric sulfur, cloud albedo and climate. Nature 326:655–661

Clilverd MA, Seppälä A, Rodger CJ, Mlynczak MG, Kozyra JU (2009) Additional stratospheric NO_x production by relativistic electron precipitation during the 2004 spring NO_x descent event. J Geophys Res 114:A04305

Comiso J (1999) Bootstrap sea ice concentrations from Nimbus-7 SMMR and DMSP SSM/I, 1979–2007. National Snow and Ice Data Center, Boulder. Digital media (1999, updated 2008)

Cox C, Munk W (1954a) Statistics of the sea surface derived from sun glitter. J Mar Res 13(N2):198–227

Cox C, Munk W (1954b) Measurement of the roughness of the sea surface from photographs of the Sun's glitter. J Opt Soc Am 44(11):838–850

Dee DP, Uppala SM, Simmons AJ, Berrisford P, Poli P, Kobayashi S, Andrae U, Balmaseda MA, Balsamo G, Bauer P, Bechtold P, Beljaars ACM, van de Berg L, Bidlot J, Bormann N, Delsol C, Dragani R, Fuentes M, Geer AJ, Haimberger L, Healy SB, Hersbach H, Hólm EV, Isaksen L, Kållberg P, Köhler M, Matricardi M, McNally AP, Monge-Sanz BM, Morcrette J–J, Park B-K, Peubey C, de Rosnay P, Tavolato C, Thépaut J-N, Vitart F (2011) The ERA-interim reanalysis: configuration and performance of the data assimilation system. Quart J R Met Soc 137:553–597

Eyring V, Köhler HW, van Aardenne J, Lauer A (2005) Emissions from international shipping: 1. The last 50 years. J Geophys Res 110:D17305

Falkowski PG, Barber RT, Smetacek V (1998) Biogeochemical controls and feedbacks on ocean primary production. Science 281:200

Fowler C (2003) Polar Pathfinder Daily 25 km EASE-grid Sea Ice Motion Vectors, 1979–2006. National snow and ice data center, Boulder. Digital media (2003, updated 2008)

Funke B, López-Puertas M, Gil-López S, von Clarmann T, Stiller GP, Fischer H, Kellmann S (2005) Downward transport of upper atmospheric NO_x into the polar stratosphere and lower mesosphere during the Antarctic 2003 and Arctic 2002/2003 winters. J Geophys Res 110:D24308

Funke B, Baumgaertner A, Calisto M, Egorova T, Jackman CH, Kieser J, Krivolutsky A, López-Puertas M, Marsh DR, Reddmann T, Rozanov E, Salmi S-M, Sinnhuber M, Stiller GP, Verronen PT, Versick S, von Clarmann T, Vyushkova TY, Wieters N, Wissing JM (2011) Composition changes after the "Halloween" solar proton event: the high energy particle precipitation in the atmosphere (HEPPA) model versus MIPAS data intercomparison study. Atmos Chem Phys 11:9089–9139

Gallet J-C, Dominé F, Arnaud L, Picard G, Savarino J (2010) Vertical profiles of the specific surface area of the snow at Dome C, Antarctica. The Cryosphere Discuss 4:1647–1708. doi:10.5194/tcd-4-1647-2010

Garcia RR, Marsh DR, Kinnison DE, Boville BA, Sassi F (2007) Simulation of secular trends in the middle atmosphere, 1950–2003. J Geophys Res 112:D09301

Gregg WW, Casey NW (2007) Modeling coccolithophores in the global oceans. Deep Sea Res Pt. II, 54(5–7), 447–477

Haltrin VI (2006) Absorption and scattering of light in natural waters. In: Kokhanovsky AA (ed) Light scattering reviews. Springer, Praxis Publishing, Chichester, pp 445–486

Hoffmann CG, Raffalski U, Palm M, Funke B, Golchert SHW, Hochschild G, Notholt J (2011) Observation of strato-mesospheric CO above Kiruna with ground-based microwave radiom-etry—retrieval and satellite comparison. Atmos Meas Tech 4:2389–2408

Hoffmann CG, Kinnison DE, Garcia RR, Palm M, Notholt J, Raffalski U, Hochschild G (2012) CO at 40-80 km above Kiruna observed by the ground-based microwave radiometer KIMRA and simulated by the whole atmosphere community climate model. Atmos Chem Phys 12:3261–3271

Holligan PM, Fernandez E, Aiken J, Balch WM, Boyd P, Burkill PH, Finch M, Groom SB, Malin G, Muller K, Purdie DA, Robinson C, Trees CC, Turner SM, Van der Wal P (1993) A biogeochemical study of the coccolithophore *Emiliania huxleyi* in the North Atlantic. Global Biogeochem Cyc 7(4):879–900

Holton JR, Haynes PH, McIntyre ME, Douglass AR, Rood RB, Pfister L (1995) Stratosphere–troposphere exchange. Rev Geophys 33:403–439

Huijnen V, Eskes HJ, Poupkou A, Elbern H, Boersma KF, Foret G, Sofiev M, Valdebenito A, Flemming J, Stein O, Gross A et al (2010) Comparison of OMI NO_2 tropospheric columns with an ensemble of global and European regional air quality models. Atmos Chem Phys 10:3273–3296

Jackman CH, DeLand MT, Labow GJ, Fleming EL, Weisenstein DK, Ko MKW, Sinnhuber M, Russell JM (2005) Neutral atmospheric influences of the solar proton events in October–November 2003. J Geophys Res 110:A09S27

Jacob DJ, Crawford JH, Kleb MM, Connors VS, Bendura RJ, Raper JL, Sachse GW, Gille JC, Emmons L, Heald CL (2003) Transport and chemical evolution over the Pacific (TRACE-P) aircraft mission: design, execution, and first results. J Geophys Res 108:9000

Konovalov IB, Beekmann M, Richter A, Burrows JP, Hilboll A (2010) Multi-annual changes of NO_x emissions in megacity regions: nonlinear trend analysis of satellite measurement based estimates. Atmos Chem Phys 10:8481–8498

Kopelevich OV (1983) Small-parameter model of optical properties of seawater. In: Monin AS (ed) Ocean optics, physical ocean optics, vol 1. Nauka, Moscow (in Russian), pp 208–234

Korb RE, Whitehouse MJ (2004) Contrasting primary production regimes around South Georgia, Southern Ocean: large blooms versus high nutrient, low chlorophyll waters. Deep-Sea Res I 51(5). doi:10.1016/j.dsr.2004.02.006

Lannuzel D, Shoemann V, de Jong J, Tison JL, Chou L (2007) Distribution and biogeochemical behaviour of iron in the East Antarctic sea ice. Mar Chem 106. doi:10.1016/j.marchem.2006.06.010

Lary DJ (1997) Catalytic destruction of stratospheric ozone. J Geophys Res 102:21515–21526

Levelt PF, van den Oord GHJ, Dobber MR, Malkki A, Huib Visser H, Johan de Vries G, Stammes P, Lundell JOV, Saari H (2006) The ozone monitoring instrument. IEEE Trans Geosci Rem Sens 44:1093–1101

Longhurst AR (1998) Ecological geography of the sea, Academic Press, San Diego, Calif

López-Puertas M, Funke B, Gil-López S, von Clarmann T, Stiller GP, Höpfner M, Kellmann S, Fischer H, Jackman CH (2005) Observation of NO_x enhancement and ozone depletion in the northern and southern hemispheres after the October–November 2003 solar proton events. J Geophys Res 110:A09S43

Lövblad G, Tarrasón L, Tørseth K, Dutchak S (2004) EMEP assessment part I European perspective. Norwegian Meteorological Institute, Oslo

Mackey MD, Mackey DJ, Higgins HW, Wright SW (1996) CHEMTAX—a program for estimating class abundances from chemical markers: application to HPLC measurements of phytoplankton. Mar Ecol Prog Ser 144:265–283

Marinov I, Gnanadesikan A, Toggweiler JR, Sarmiento JL (2006) The Southern Ocean biogeochemical divide. Nature 441. doi:10.1038/nature04883

Milliman JD (1993) Production and accumulation of calcium in the ocean. Global Biogeochem Cyc 7:927–957

Mobley CD (1994) Light and water. Radiative transfer in natural waters. Academic Press, San Diego

Morel A (1974) Optical properties of pure water and pure seawater. In: Jerlov NG, Steemann Nielsen E (eds) Optical aspects of oceanography. Academic, New York, pp 1–24

Morel A, Maritorena S (2001) Bio-optical properties of oceanic waters: a reappraisal. J Geophys Res 106(C4):7163–7180

Mudelsee M (2010) Climate time series analysis. Springer, Dordrecht

Nair A, Sathyendranath S, Platt T, Morales J, Stuart V, Forget M, Devred E, Bouman H (2008) Remote sensing of phytoplankton functional types. Remote Sens Environ 112:3366–3375

Nolin A, Dozier J (2000) A hyperspectral method for remotely sensing the grain size of snow. Remote Sens Environ 74:207–216. doi:10.1016/S0034-4257(00)00111-5

Nüß JH (2005) Improvements of the retrieval of tropospheric NO$_2$ from GOME and SCIAMACHY data. Ph D thesis, University of Bremen

O'Reilly JE, Maritorena S, Mitchell BG, Siegel DA, Carder KL, Garver SA, Kahru M, McClain C (1998) Color chlorophyll algorithms for SeaWiFS. J Geophys Res 103(C11):24937–24953

Painter T, Molotch N, Cassidy M, Flanner M, Steffen K (2007) Contact spectroscopy for determination of stratigraphy of snow optical grain size. J Glaciol 53:121–127. doi:10.3189/172756507781833947

Park J, Oh I-S, Kim H-C, Yoo S (2010) Variability of SeaWiFS chlorophyll-a in the southwest Atlantic sector of the Southern Ocean: strong topographic effects and weak seasonality. Deep-Sea Res I 57. doi:10.1016/j.dsr.2010.01.004

Parkinson CL, Cavalieri DJ (2002) A 21 year record of Antarctic sea-ice extents and their regional, seasonal and monthly variability and trends. Ann Glac 34:441–446

Perner D, Platt U (1979) Detection of nitrous acid in the atmosphere by differential optical absorption. Geophys Res Lett 93:917–920

Platt U, Stutz J (2008) Differential optical absorption spectroscopy. Springer, Berlin

Pope RM, Fry ES (1994) Absorption spectrum (380–700 nm) of pure water. II. Integrating cavity measurements. Appl Optics 36(33):8710–8723

Porter HS, Jackman CH, Green AES (1976) Efficiencies for production of atomic nitrogen and oxygen by relativistic proton impact in air. J Chem Phys 65:154–167

Powell MJD (1964) An efficient method for finding the minimum of a function of several variables without calculating derivatives. Comput J 7(2):155–162

Prieur L, Sathyendranath S (1981) An optical classification of coastal and oceanic waters based on the specific absorption curves of phytoplankton pigments, dissolved organic matter, and other particulate materials. Limnol Oceanogr 26:671–689

Raitsos DE, Lavender SJ, Pradhan Y, Tyrrell T, Reid PC, Edwards M (2006) Coccolithophore bloom size variation in response to the regional environment of the subarctic North Atlantic. Limnol Oceanogr 51:2122–2130

Randall D, Wood R, Bony S, Colman R, Fichefet T, Fyfe J, Kattsov V, Pitman A, Shukla J, Srinivasan J, Stouer R, Sumi A, Taylor K (2007a) Climate models and their evaluation. In: Solomon S, Qin D, Manning M, Chen Z, Marquis M, Averyt K, Tignor M, Miller H (eds) Climate change 2007: the physical science basis. Contribution of working group I to the fourth assessment report of the intergovernmental panel on climate change. Cambridge University Press, Cambridge

Randall CE, Harvey VL, Singleton CS, Bailey SM, Bernath PF, Codrescu M, Nakajima H, Russell JM III (2007b) Energetic particle precipitation effects on the Southern hemisphere stratosphere in 1992–2005. J Geophys Res 112:D08308

Randall CE, Harvey VL, Siskind DE, France J, Bernath PF, Boone CD, Walker KA (2009) NO$_x$ descent in the Arctic middle atmosphere in early 2009. Geophys Res Lett 36:L18811

Rao KN, Weber A (1992) Spectroscopy of the Earth's atmosphere and interstellar medium. Academic Press, Boston

Richter A, Burrows JP, Nusz H, Granier C, Niemeier U (2005) Increase in tropospheric nitrogen dioxide over China observed from space. Nature 437:129–132

Ridder T, Gerbig C, Notholt J, Rex M, Schrems O, Warneke T, Zhang L (2011) Ship-borne FTIR measurements of CO and O$_3$ in the western pacific from 43°N to 35°S: an evaluation of the sources. Atmos Chem Phys 12:815–828

Röttgers R, Haese C, Dörffer R (2007) Determination of the particulate absorption of microalgae using a point-source integrating-cavity absorption meter: verification with a photometric technique, improvements for pigment bleaching, and correction for chl. fluorescence. Limnol Oceanogr Methods 5:1–12

Rozanov A (2008) SCIATRAN 2.X: radiative transfer model and retrieval software package. URL = http://www.iup.physik.uni-bremen.de/sciatran

Rozanov VV, Buchwitz M, Eichmann K-U, de Beek R, Burrows JP (2002) Sciatran—a new radiative transfer model for geophysical applications in the 240–2400 nm spectral region: the pseudo-spherical version. Adv Space Res 29:1831–1835

Rozanov A, Rozanov VV, Buchwitz M, Kokhanovsky A, Burrows JP (2005) SCIATRAN 2.0—a new radiative transfer model for geophysical applications in the 175–2400 nm spectral region. Adv Space Res 36:1015–1019

Rusch DW, Gerard JC, Solomon S, Crutzen PJ, Reid GC (1981) The effect of particle precipitation events on the neutral and ion chemistry of the middle atmosphere—I Odd nitrogen. Planet Space Sci 29:767–774

Sarmiento JL, Gruber N (2006) Ocean biogeochemical dynamics. Princeton University Press, Princeton

Sathyendranath S, Watts L, Devred E, Platt T, Caverhill C, Maass H (2004) Discrimination of diatoms from other phytoplankton using ocean-colour data. Mar Ecol Prog Ser 272:59–68

Scambos T, Haran T, Fahnestock M, Painter T, Bohlander J (2007) MODIS-based mosaic of Antarctica (MOA) data sets: continent-wide surface morphology and snow grain size. Remote Sens Environ 111:242–257. doi:10.1016/j.rse.2006.12.020

Schlitzer R (2002) Carbon export fluxes in the Southern Ocean: results from inverse modeling and comparison with satellite-based estimates, Deep-Sea Res II 49. doi:10.1016/S0967-0645(02)00004-8

Schwegmann S, Haas C, Fowler C, Gerdes R (2011) A comparison of satellite-derived sea-ice motion with drifting-buoy data in the Weddell Sea, Antarctica. Ann Glac 52(57):103–110

Seinfeld JH, Pandis SN (2006) Atmospheric chemistry and physics: from air pollution to climate change. 2nd edn. Wiley, Hoboken

Seppälä A, Verronen PT, Kyrölä E, Hassinen S, Backman L, Hauchecorne A, Bertaux JL, Fussen D (2004) Solar proton events of October–November 2003: ozone depletion in the northern hemisphere polar winter as seen by GOMOS/Envisat. Geophys Res Lett 31:L19107

Shchepetkin AF, McWilliams JC (2005) The regional oceanic modeling system (ROMS): a split-explicit, free-surface, topography-following-coordinate oceanic model. Ocean Mod 9. doi:10.1016/j.ocemod.2004.08.002

Shifrin KS (1988) Physical optics of ocean water. AIP translation series. Amer Inst Phys N Y 285

Sinnhuber B-M, Weber M, Amankwah A, Burrows JP (2003a) Total ozone during the unusual Antarctic winter of 2002. Geophys Res Lett 30(11):1580–1584

Sinnhuber M, Burrows JP, Chipperfield MP, Jackman CH, Kallenrode M-B, Künzi KF, Quack M (2003b) A model study of the impact of magnetic field structure on atmospheric composition during solar proton events. Geophys Res Lett 30:1818–1821

Sinnhuber B-M, Weber M, Amankwah A, Burrows JP (2003c) Total ozone during the unusual Antarctic winter of 2002. Geophys Res Lett 30:1580–1583

Sinnhuber M, Kazeminejad S, Wissing JM (2011) Interannual variation of NO_x from the lower thermosphere to the upper stratosphere in the years 1991–2005. J Geophys Res 116:A02312

Smith KR (1993) Fuel combustion, air pollution exposure, and health: the situation in developing countries. Annu Rev Energy Environ 18:529–566

Solomon S, Rusch D, Gerard J, Reid G, Crutzen P (1981) The effect of particle-precipitation events on the neutral and ion chemistry of the middle atmosphere—2 Odd hydrogen. Planet Space Sci 29:885–892

Solomon S, Garcia RR, Olivero JJ, Bevilacqua RM, Schwartz PR, Clancy RT, Muhleman DO (1985) Photochemistry and transport of carbon monoxide in the middle atmosphere. J Atmos Sci 42:1072–1083

Solomon S, Qin D, Manning M, Chen Z, Marquis M, Averyt KB, Tignor M, Miller HL (2007) Contribution of working group I to the fourth assessment report of the intergovernmental panel on climate change. Cambridge University Press, Cambridge

Stamnes K, Li W, Eide H, Aoki T, Hori M, Storvold R (2007) ADEOSII/GLI snow/ice products—part I: scientific basis. Remote Sens Environ 111:258–273. doi:10.1016/j.rse.2007.03.023

Stroeve JC, Serreze MC, Fetterer F, Arbetter T, Meier W, Maslanik J, Knowles K (2005) Tracking the Arctic's shrinking ice cover: another extreme September minimum in 2004. Geophys Res Lett 32. doi:10.1029/2004GL021810

Swider W, Keneshea TJ (1973) Decrease of ozone and atomic oxygen in lower mesosphere during a PCA event. Planet Space Sci 21:1969–1973

Tagliabue A, Bopp L, Aumont O (2009) Evaluating the importance of atmospheric and sedimentary iron sources to Southern Ocean biogeochemistry. Geophys Res Lett 36. doi:10.1029/2009GL038914

Thompson DWJ, Solomon S (2002) Interpretation of recent Southern hemisphere climate change. Science 296:895–899

Timmermann R, Danilov S, Schröter J, Böning C, Sidorenko D, Rollenhagen K (2009) Ocean circulation and sea ice distribution in a finite element global sea ice-ocean model. Ocean Model. doi:10.1016/j.ocemod.2008.10.009

Tyrrell T, Holligan PM, Mobley CD (1999) Optical impacts of oceanic coccolithophore blooms. J Geophys Res 104(C2):3223–3241

U.S. Geological Survey: Global Digital Elevation Model (GTOPO30) (2004)

United Nations Department of Economic and Social Affairs (2010) World urbanization prospects—the 2009 revision: highlights. United Nations Department of Economic and Social Affairs, New York

van der ARJ, Eskes HJ, Boersma KF, van Noije TPC, Roozendael MV, Smedt ID, Peters DHMU, Meijer EW (2008) Trends, seasonal variability and dominant NO_x source derived from a ten year record of NO_2 measured from space. J Geophys Res 113:D04302

Verronen PT, Funke B, López-Puertas M, Stiller GP, von Clarmann T, Glatthor N, Enell C-F, Turunen E, Tamminen J (2008) About the increase of HNO_3 in the stratopause region during the Halloween 2003 solar proton event. Geophys Res Lett 35:L20809

Vountas M, Dinter T, Bracher A, Burrows JP, Sierk B (2007) Spectral studies of ocean water with space-borne sensor SCIAMACHY using differential optical absorption spectroscopy (DOAS). Ocean Sci 3:429–440

Wang P, Stammes P, van der AR, Pinardi G, van Roozendael M (2008) FRESCO+: an improved O_2 A-band cloud retrieval algorithm for tropospheric trace gas retrievals. Atmos Chem Phys 8:6565–6576

Westbroek P, De Vring-De Jong EW, Van Der Wal P, Borman AH, De Vring JPM (1985) Biopolymer-mediated Ca and Mn accumulation and biomineralization. Geol Mijnbouw 64:5–15

Wiebe H (2011) Implementation and validation of the snow grain size retrieval SGSP from spectral reflectances of the satellite sensor MODIS. Ph D thesis, p 106

Wiebe H, Heygster G, Zege E (2011) Snow grain size retrieval SGSP from optical satellite data: validation with ground measurements and detection of snow fall events. Rem Sens Environ (in press)

Winder M, Cloernet JE (2010) The annual cycles of phytoplankton biomass. Phil Trans R Soc 365:3215–3226

Winkler H, Sinnhuber M, Notholt J, Kallenrode M-B, Steinhilber F, Vogt J, Zieger B, Glassmeier K-H, Stadelmann A (2008) Modeling impacts of geomagnetic field variations on middle atmospheric ozone responses to solar proton events on long timescales. J Geophys Res 113:D02302

Winkler H, Kazeminejad S, Sinnhuber M, Kallenrode M-B, Notholt J (2009) Conversion of mesospheric HCl into active chlorine during the solar proton event in July 2000 in the northern polar region. J Geophys Res 114:D00I03

Wissing JM, Kallenrode M-B (2009) Atmospheric Ionization Module Osnabrück (AIMOS): a 3-D model to determine atmospheric ionization by energetic charged particles from different populations. J Geophys Res 114:A06104

Wissing JM, Kallenrode M-B, Wieters N, Winkler H, Sinnhuber M (2010) Atmospheric ionization module Osnabrück (AIMOS): 2. Total particle inventory in the October–November 2003 event and ozone. J Geophys Res 115:A02308

WMO (World Meteorological Organization) (2007) Scientific assessment of ozone depletion: 2006, global ozone research and monitoring project—report no. 50. Geneva

Yoder JA, McClain CR, Feldman GC, Esaias WE (1993) Annual cycles of phytoplankton chlorophyll concentrations in the global ocean: a satellite view. Global Biogeochem Cycles 7(1):181–193

Zege E, Kokhanovsky A, Katsev I, Polonsky I, Prikhach A (1998) The retrieval of the effective radius of snow grains and control of snow pollution with GLI data. In: Mishchenko M, Travis L, Hovenier J (eds) Proceedings of conference on light scattering by nonspherical particles: theory, measurements, and applications. American Meteorological Society, Boston, pp 288–290

Zege E, Katsev I, Malinka A, Prikhach A, Polonsky I (2008) New algorithm to retrieve the effective snow grain size and pollution amount from satellite data. Ann Glaciol 49:139–144. doi:10.3189/172756408787815004

Zege EP, Katsev IL, Malinka AV, Prikhach AS, Heygster G, Wiebe H (2011) Algorithm for retrieval of the effective snow grain size and pollution amount from satellite measurements. Remote Sens Environ 115(10):2674–2685

Zhang Q, Streets DG, He K, Wang Y, Richter A, Burrows JP, Uno I, Jang CJ, Chen D, Yao Z, Lei Y (2007) NO$_x$ emission trends for China, 1995–2004: the view from the ground and the view from space. J Geophys Res 112:D22306

Chapter 3
Earth System Modelling and Data Analysis

3.1 The Last Interglacial as Simulated by an Atmosphere–Ocean General Circulation Model: Sensitivity Studies on the Influence of the Greenland Ice Sheet

Madlene Pfeiffer (✉) and Gerrit Lohmann

Alfred Wegener Institute for Polar and Marine Research, Bremerhaven, Germany
e-mail: madlene.pfeiffer@awi.de

Abstract During the Last Interglacial (LIG), the northern high latitudes showed summer temperatures higher than those of the late Holocene, and a significantly reduced Greenland Ice Sheet (GIS). We perform sensitivity studies for the height and extent of the GIS at the beginning of the LIG [130 kyr before present (BP)], using the COSMOS coupled atmosphere–ocean general circulation model. Different methods are deployed in order to change the GIS height and surface area. Our experimental approach also considers the Earth's orbital parameters for 130 kyr BP, since insolation changes are considered to be the main driver of LIG warmth. We analyze resulting anomalies in surface air temperature and sea ice cover. Our study shows that a strong Northern Hemisphere warming indeed is mainly caused by increased summer insolation. Changes in GIS elevation, surface area, and albedo contribute to the overall warming of the LIG, but any of these changed model boundary conditions lead to a weaker effect than the adjusted orbital forcing.

Keywords The last interglacial · Preindustrial · Greenland ice sheet · Surface temperature · Sea ice · Orbital forcing

G. Lohmann et al. (eds.), *Earth System Science: Bridging the Gaps between Disciplines*, SpringerBriefs in Earth System Sciences,
DOI: 10.1007/978-3-642-32235-8_3, © The Author(s) 2013

3.1.1 Introduction

Climate models represent an important tool in understanding the complex climate of the Earth. Coupled atmosphere–ocean general circulation models (AOGCM) simulate atmospheric and oceanic physical processes, and the interaction between them. One important application of such models is the simulation of future climate scenarios (Jansen et al. 2007). Obviously, a correct prediction of the climate of the future has importance for the human society as a whole. To ensure reliability of climate predictions, the models need to be tested on their ability to sufficiently reproduce mean climate states that are different from today. Past geologic time-scales are useful for this purpose. Simulations of interglacial climates provide the possibility to evaluate how models respond when strong changes in the forcing are applied (Mearns et al. 2001).

In our study, we discuss results from AOGCM simulations of the beginning of the LIG at 130 kyr before present (BP). The LIG represents the interglacial before the Holocene, which itself represents a warm stage. The LIG climate is considered to have been warmer than the Holocene. It is believed that one cause for this difference is increased LIG insolation during the summer months at mid-to-high latitudes, while winter insolation has been lower in the tropics. The enhanced seasonality in the Northern Hemisphere (NH) is thought to be caused by larger obliquity (ε), eccentricity (e), and angle of perihelion (ω) of the Earth's orbit (Berger 1978).

During the LIG Arctic regions experienced summer temperatures warmer than those of the late Holocene, and the Greenland Ice Sheet (GIS) has been significantly reduced (Otto-Bliesner et al. 2006). Here, we investigate the relative contribution of three factors on LIG warmth: orbital forcing, the reduced elevation and extent of the GIS, and albedo changes. To this end, we perform four sensitivity studies, which include not only the orbital forcing for 130 kyr BP, but also variations of the GIS.

3.1.2 Model Description and Experimental Setup

3.1.2.1 Model Description

The fully coupled Community Earth System Models (COSMOS) consist of the atmosphere model ECHAM5 (Roeckner et al. 2003), the land surface model JSBACH (Raddatz et al. 2007), the general ocean circulation model MPIOM (Marsland et al. 2003), and the OASIS3 coupler (Valcke et al. 2003). COSMOS is being developed at the Max-Planck-Institute for Meteorology in Hamburg and other institutions. The atmospheric component ECHAM5 is a spectral model with a horizontal resolution of T31 ($\sim 3.75° \times 3.75°$). It has 19 vertical hybrid sigma-pressure levels, the highest level being located at 10 hPa. The land surface model

JSBACH simulates water, fluxes of energy, momentum and CO_2 between land an atmosphere, and comprises a dynamic vegetation module (Brovkin et al. 2009) which enables the vegetation to explicitly adjust to a change in the climate state. MPIOM has a resolution of GR30 (corresponding to $\sim 3°$) with 40 vertical levels on an Arakawa C-grid bipolar orthogonal spherical coordinate system. It includes a Hibler-type zero-layer dynamic-thermodynamic sea ice model with viscous plastic rheology (Semtner 1976; Hibler 1979). MPIOM does not apply a flux correction (Jungclaus et al. 2006), allowing for a better representation of oceanic heat transports in comparison to flux-corrected models. The time steps are 40 min in the atmosphere and 144 min in the ocean.

3.1.2.2 Experimental Setup

We use a preindustrial (PI) simulation (Wei et al. 2012) as a control climate. The greenhouse gas concentrations and orbital forcing of this PI experiment are pre-scribed according to the Paleoclimate Modelling Intercomparison Project Phase 2 protocol (Braconnot et al. 2007a, b). The simulations of the LIG are performed as equilibrium experiments with fixed boundary conditions. Orbital parameters for this time slice have been calculated according to Berger (1978) for 130 kyr BP ($e = 0.038231$, $\varepsilon = 24.2441°$, and $\omega = 49.097°$). Our main interest is in the effects of orbital forcing and the shape of the GIS, therefore greenhouse gas concentrations have been prescribed at PI levels (278 ppmv CO_2, 650 ppbv CH_4, and 270 ppbv N_2O).

The extent to which the GIS has been reduced during the LIG is not well constrained by observations. Therefore, we perform four experiments that differ in Greenland elevation and ice sheet area. An LIG control run (LIG-ctl) with present GIS allows us to quantify the exclusive effects of orbital forcing. Three experiments consider a modified GIS: (1) GIS lowered to half its present elevation (LIG- × 0.5), but unchanged GIS area; (2) GIS lowered by a maximum of 1300 m which is about half of its present elevation (LIG-1300m). At locations where the present Greenland elevation is less than 1300 m, we set orography to zero meters, define the ground to be ice-free, but keep the albedo at values typical for the GIS. (3) Similar to LIG-1300m, but introduced ice-free areas also include a change in the background albedo (LIG-1300m-alb). This experimental approach allows us to separate the effects of a lowered and shrunk GIS from those of changes in albedo on the climate of the NH. Other boundary conditions of the LIG experiments are kept constant at their PI state. We integrated every experiment for 500 model years into a quasi-equilibrium. Only the last 50 years of the integration are used for our analysis.

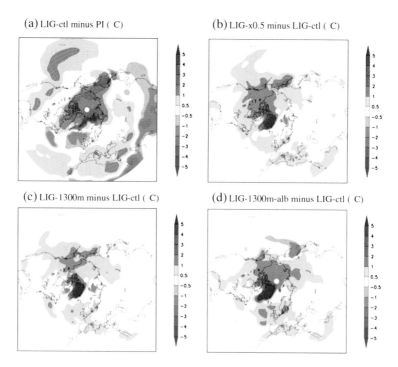

(a) LIG-ctl minus PI (C)   **(b)** LIG-x0.5 minus LIG-ctl (C)

(c) LIG-1300m minus LIG-ctl (C)   **(d)** LIG-1300m-alb minus LIG-ctl (C)

Fig. 3.1 Annual mean surface temperature anomalies north of 20°N for: **a** LIG-ctl minus PI, **b** LIG-×0.5 minus LIG-ctl, **c** LIG-1300m minus LIG-ctl, **d** LIG-1300m-alb minus LIG-ctl

3.1.3 Results

We focus on anomalies of surface air temperature (SAT) and sea ice as important climate indicators of the NH (Figs. 3.1, 3.2). In Figs. 3.1a, 3.2a we see the effect of the orbital forcing. Annual mean SAT north of 20°N in the LIG is predominantly higher than during the PI (ΔTS = +0.47 °C). LIG northern high latitudes (60–90°N) are significantly warmer (ΔTS = +1.39 °C, regionally up to 7.5 °C) relative to PI. The Arctic Ocean is warmed by around 2–3 °C. Strongest warming occurs over the Barents and the Greenland Seas, where we observe an increased cloud-cover in the LIG (not shown here). Cooling occurs at $\sim$20°N over Africa, at $\sim$25°N over the Arabian Peninsula and India. We observe a link between the SAT and sea ice cover in the northern high latitudes, which show similar anomaly patterns (Fig. 3.2a). Sea ice cover in the LIG is generally much reduced.

Figures 3.1b, 3.2b present the effect of lowering the GIS by half its present elevation. We observe a strong warming relative to PI over Greenland of up to 11.1 °C, and a warming of $\sim$2 °C over Canada, Alaska, and the western Arctic Ocean. The Bering Sea warms by up to 3 °C, while Russia and the eastern part of the Arctic Ocean warm by up to 1 °C. The mean temperature anomaly north of 20°N is ΔTS = +0.56 °C, in the northern high latitudes it amounts to

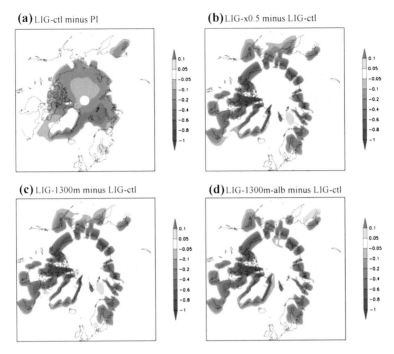

Fig. 3.2 Annual mean sea ice cover anomalies north of 50°N for: **a** LIG-ctl minus PI, **b** LIG-×0.5 minus LIG-ctl, **c** LIG-1300m minus LIG-ctl, **d** LIG-1300m-alb minus LIG-ctl

ΔTS = +1.07 °C. Cooling due to a lowering of the GIS is limited to the Barents Sea with anomalies of ΔTS = −1.6 °C. Temperature anomalies coincide with variations in the sea ice pattern with decreasing (increasing) sea ice cover in warming (cooling) areas, e.g. in the Barents Sea.

Figures 3.1c, 3.2c depict the effects of lowering the GIS by 1,300 m of its present elevation, while retaining the background albedo. Average temperature anomalies above 20°N and in northern high latitudes are ΔTS = +0.45 °C and ΔTS = +1.03 °C, respectively. In the Barents Sea and south-west of Greenland, we observe a cooling that reaches ΔTS = −1.6 °C, while other regions warm. As in Figs. 3.1a, b and 3.2a, b, we find a close connection between changes in SAT and sea ice cover.

In Figs. 3.1d, 3.2d we observe the effect of lowering GIS by 1,300 m of its present elevation including albedo changes. Compared to Figs. 3.1c, 3.2c, we find a warming that amounts north of 20°N to ΔTS = +0.51 °C, and north of 60°N to ΔTS = +1.45 °C. We observe a cooling over the Sea of Okhotsk and a warming over other areas. There is again a close link between anomalies in SAT and sea ice cover.

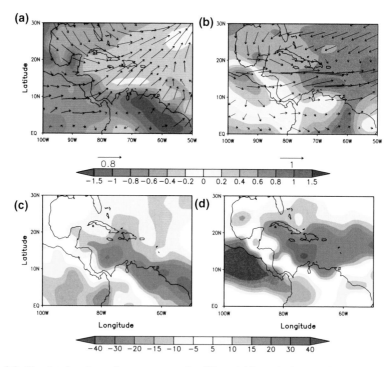

Fig. 3.3 Simulated surface air temperature (in °C) and 10 m wind anomalies (m/s) in H6 K relative to CTL in the Caribbean for **a** March–May (MAM) and **b** September–December (SON). **c** and **d** as **a** and **b**, but for precipitation anomalies (in mm/mon). Here, the anomalies in seasons with minimum and maximum values are shown instead of conventional summer and winter seasons

3.1.4 Discussion and Conclusions

It is known that the main trigger for LIG warmth is the change in the Earth's orbital parameters (Kutzbach et al. 1991; Crowley and Kim 1994; Montoya et al. 2000; Felis et al. 2004; Kaspar and Cubasch 2007). In general, the NH is warming in our LIG experiments. We can confirm the importance of insolation for the NH, and especially for the northern high latitudes (Figs. 3.3a, 3.4a). There is a regional cooling over the North of Africa, the Arabian Peninsula, and India. This effect has been observed before by Herold and Lohmann (2009), who also propose a mechanism for the temperature anomalies which relies on changes in insolation in conjunction with cloud cover and zonal winds.

In all GIS sensitivity studies we observe widespread warming in the NH. The strong rise of temperatures above Greenland is directly related to the lapse rate via reduction of the ice sheet elevation to half its present value. There are rather small changes in atmospheric circulation in the northern high latitudes. The prevailing easterlies therefore transport air, which has been heated up above Greenland,

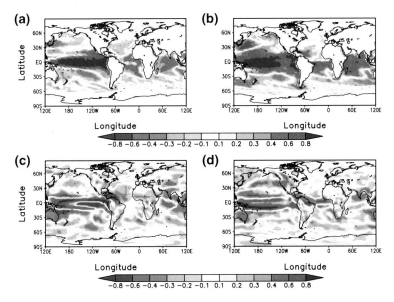

Fig. 3.4 Correlation maps of simulated Nino3 index with SST during the last 100 years of the integration in CTL for **a** September–November (0) and **b** March–May (+1). **c** and **d** as **a** and **b**, but for precipitation. Here, Nino3 index is defined as the December (0)–February (+1) mean SST anomalies over the area 150–90° W, −5–5° N. Significance greater than 95 % is plotted

towards Canada, Alaska, the western Arctic Ocean and the Bering Sea. As a result, the sea ice cover is much lower during the LIG compared to PI.

Our experiments allow us to quantify the relative contribution of orbital forcing, and of changes in the size of the GIS, and the related albedo. North of 20°N, the combined effects of insolation and elevation changes of the GIS during the LIG (LIG-×0.5 minus PI) amount to an average warming of +1.03 °C. This equals the sum of the average warming caused by insolation alone (LIG-ctl minus PI, +0.47 °C), and the warming caused by lowering of the GIS (LIG-×0.5 minus LIG-ctl, +0.56 °C) in this area. Including the effects of albedo (LIG-1300m-alb minus PI), leads to a warming of +0.98 °C. Changes in albedo alone lead to a warming of +0.06 °C. North of 60°N, the combined effects of insolation and GIS elevation changes amount to a warming of +2.46 °C. The orbital forcing leads to a warming of +1.39 °C, while the elevation changes warm the northern high latitudes by +1.07 °C. Combined changes of insolation, albedo, and decreased GIS elevation cause an overall warming of +2.84 °C. The exclusive effect of albedo changes in the northern high latitudes amounts to +0.42 °C. Therefore we find that the warming effect of insolation and GIS changes indeed is strongly focused on the high latitudes of the NH. As expected, a combination of all contributing effects leads to the highest warming north of 60°N. Interestingly, this is not the case if we average the temperature anomaly north of 20°N.

We are aware of only one other model study of the LIG that includes changes in GIS. It has been presented by Otto-Bliesner et al. (2006). They performed a similar

GIS sensitivity experiment, but assumed a reduction of the GIS elevation of about 500 m. In their results, they show that the JJA temperature anomaly with respect to PI is positive in the NH especially over the continents. The warming we find in our study is even more pronounced due to the stronger lowered GIS elevation, but the pattern is similar (not shown here).

We conclude that our AOGCM reproduces an LIG climate that is comparable to other studies. We find that the LIG is significantly warmer than the PI in the NH north of 20°N. These changes are predominantly caused by an increase in insolation, followed by the effect of a lowered GIS, and subsequent changes in albedo. These sensitivity studies represent the starting point for transient integrations of the Eemian climate and comparison with high-quality proxy data (Drysdale et al. 2009).

3.2 Simulated Caribbean Climate Variability During the Mid-Holocene

Wei Wei (✉) and Gerrit Lohmann

Alfred Wegener Institute for Polar and Marine Research, Bremerhaven, Germany
e-mail: wei.wei@awi.de

Abstract The external and internal induced climate variability in the Caribbean during the mid-Holocene is investigated using long-term integration of the Earth system model COSMOS in this study. The orbital induced insolation change is the primary forcing controlling the climatology change during the mid-Holocene, resulting in an increased seasonality and much wetter condition in this region. The seasonal and interannual variability associated with the El Niño-Southern Oscillation (ENSO) and their influence on the Caribbean climate show no significant change during the mid-Holocene. The Atlantic Multidecadal Oscillation is the forcing mechanism for the decadal to centennial variability in the Caribbean. A warming and wetter condition in the Caribbean follows the positive AMO phase in both present-day and mid-Holocene, implying that it is a stable internal pattern in the climate system during the Holocene. We suggest that the results can be favorable for better understanding of the Caribbean climate change in the future.

Keywords Caribbean ·Mid-holocene · Orbital forcing · Earth system models COSMOS · Seasonality · Climate variability · El Niño-Southern oscillation · Atlantic multidecadal oscillation

3.2.1 Introduction

The climate variability from the mid-Holocene (6,000 years before present) to the present is of particular importance to understand the past climate change, due to the relative stable boundary conditions in this period. The orbital change is the major external forcing and can have considerable influence on the Northern Hemisphere mid-high latitude climate. Comparably, its influence on the tropical and subtropical regions is moderate. Concerning the Caribbean region, this influence is even minor, with the annual mean temperature anomalies much smaller than 1 °C during the mid-Holocene relative to the present (Braconnot et al. 2007a, b). Results based on the proxy studies (e.g. Giry et al. 2012) reveal that there is a cooling trend during the Holocene in the Caribbean, but its magnitude is still controversial.

On seasonal and interannual timescales, the Caribbean climate variability under the present condition is thought to be controlled mainly by the internal variability of El Niño-Southern Oscillation (ENSO; e.g. Giannini et al. 2000). On longer timescales, i.e. decadal and centennial, the Atlantic Multidecadal Oscillation play the most important role (AMO; e.g. Sutton and Hodson 2005). Studies based on proxy and modelling find a reduced ENSO activity during the mid-Holocene (e.g. Clement et al. 2000). Based on multi-proxy analysis, a quasi-persistent ~55- to 70-year AMO is found to exist during the last 8,000 years (Knudsen et al. 2011). However, the climate influence of them during the mid-Holocene has not been investigated.

Considering the relative small quantity of high resolution proxy data in the Caribbean during the Holocene, the paleo-modelling effort for this specific region is even less. In this study, we will present the simulated climatology change in the Caribbean during the Holocene based on modelling study. The forcing mechanisms controlling the Caribbean climate variability on different timescales will also be examined. Such achievements will enable us to better predict the future climate change in the Caribbean, where it is vulnerable to natural phenomena and under high development pressure.

3.2.2 Method

3.2.2.1 Model Description

Numerical experiments are performed with the Earth system model COSMOS developed by the Max-Planck-Institute for Meteorology that describes the dynamics of the atmosphere–ocean-sea ice-vegetation system. The atmospheric component is the spectral atmosphere model ECHAM5 (Roeckner et al. 2003) with the resolution of T31, corresponding to $3.75 \times 3.75°$ in horizontal, and 19 hybrid sigma pressure level in vertical. The land processes are integrated into the

Table 3.1 Boundary conditions used in the two long-term simulations

Experiment	Boundary conditions				Integration time (year)
	Orbital	Greenhouse gases			
		CO_2 (ppm)	CH_4 (ppb)	N_2O (ppb)	
CTL	Present-day	280	760	270	3000
H6 K	6,000 years BP	280	650	270	3000

atmosphere model using the land surface model JSBACH (Raddatz et al. 2007) except for river routing, for which the hydrological discharge model is responsible (Hagemann and Gates 2003). The ocean-sea ice component is the ocean general circulation model MPI-OM (Marsland et al. 2003) with the resolution of GR30 in horizontal and 40 unevenly spaced vertical levels, which includes the dynamics of sea ice formulated using viscous-plastic rheology (Hibler III 1979). The atmosphere and the ocean components interact through the OASIS3 coupler (Valcke 2006).

3.2.2.2 Experiment Setup

We have carried out two long-term experiments: a pre-industrial control experiment (CTL) and a mid-Holocene (H6 K), by prescribing the appropriate orbital parameters and greenhouse gases. For both experiments, the boundary conditions (Table 3.1) are the same as those used in Paleoclimate Modelling Intercomparison Project (PMIP) (Crucifix et al. 2005). Each experiment is run for 1,000 years as spin-up and further integrated for 2,000 years. More detailed information about the experiments setup has been elaborated in Wei et al. (2012).

3.2.3 Results

3.2.3.1 Mean Climatology

The simulated surface air temperature in the Caribbean shows a general warming in summer and autumn and a cooling in the other seasons during the mid-Holocene (Fig. 3.3a, b). Their spatial patterns and magnitudes are in general consistent with a previous Holocene transient simulation using a similar model setup (Lorenz and Lohmann 2004). Such change in the temperature leads to mid-Holocene seasonality increased by approximately 1 °C in the Caribbean, which is more than 20 % of the total seasonality. This result is also supported by the coral-based reconstruction (Giry et al. 2012).

The easterly winds show a significant decrease over the Caribbean during the mid-Holocene in all seasons, especially during the summer (Fig. 3.3a, b). These anomalies tend to result in more convergence in the Caribbean and less water vapor transported to the Western Pacific. This partly explains the remarkable increase of the Caribbean precipitation in H6 K (Fig. 3.3c, d). However, the precipitation anomalies are mainly related to the orbital induced intertropical convergence zone (ITCZ) shift. Evidences for a more northward position of the ITCZ over the Atlantic during the Holocene have been found in some proxies (e.g. Haug et al. 2001). The ITCZ shift can be seen from the simulation results, which illustrate a precipitation anomaly contrast between the northern South America and the Caribbean. Such a shift brings more precipitation to the Caribbean in all seasons (Fig. 3.3c, d). The warmer summer surface temperature (Fig. 3.3b) could provide more energy for this wetter condition by more convection.

3.2.3.2 Climate Variability Associated with the ENSO

The relationship between the seasonal to interannual climate variability in the Caribbean with the ENSO is shown by correlation maps (Fig. 3.4). During the pre-mature phase of a warm ENSO [September–November (0)], the sea surface temperature (SST) manifests positive anomalies over the tropical Pacific Ocean (Fig. 3.4a). A zonal seesaw pattern in the sea level pressure exists between the Pacific and Atlantic, with high pressure anomalies dominating the Caribbean region, which favors less precipitation (Fig. 3.4c) due to a direct atmospheric influence. With the easterly winds weakened (not shown), this warm anomalies gradually propagate eastward and can reach the Caribbean during the mature phase [December (0)–February (+1)]. After the warm ENSO, its influence can stay over the Caribbean for another season [March–May (+1)], during which the warmer SST spreads over the whole Caribbean region and also extends southward to the tropical Atlantic Ocean (Fig. 3.4b). Precipitation is still anti-correlated with the Nino3 index over the most Caribbean, especially the southern part (Fig. 3.4d).

To detect possible change of the ENSO and its influence on the Caribbean climate during the Holocene, we apply the same analysis to the mid-Holocene H6 K experiments. Interestingly, the spatial patterns of the correlation between the Nino3 indices and the SSTs have no considerable change in H6 K, which also leads to a similar pattern in the correlation maps of Nino3 indices with the precipitation in the Caribbean (not shown). This result indicates that the ENSO phenomenon can be a stable forcing factor that controls the seasonal and inter-annual climate variability in the Caribbean during the Holocene.

3.2.3.3 Climate Variability Associated with the AMO

The detailed analyses of the simulated AMO during the Holocene is conducted by Wei and Lohmann (2012). Here, we concentrate on its climate influence on the Caribbean climate. During a warm phase of the AMO, the Caribbean experiences a warming up to 0.2 °C in the simulations (Fig. 3.5a, b). Considering the relative

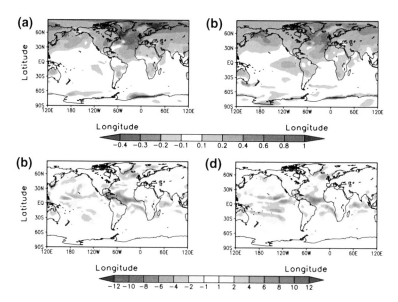

Fig. 3.5 Composite maps of surface temperature (in °C) with the AMO index for **a** CTL and **b** H6 K. **c** and **d** as **a** and **b**, but for the precipitation. The AMO indices are constructed based on simulated sea surface temperature veraged over the region 0–60°N, 7.5–75°W. The band-pass filter has been applied to both the indices and fields to assure only variations in multidecadal timescale are represented. A linear trend is also removed before the composite analyses. The composite maps shown here are calculated by subtracting the fields that have higher than one standard deviation of the mean from those with lower than one standard deviation with respect to the indices. Here anomalies with significance higher than 95 % are plotted

low seasonal and year-to-year variation in this region, the warming induced by the positive AMO is significant. Associated with the warming, the precipitation in the Caribbean is also increased considerably (Fig. 3.5c, d). The AMO influence on the Caribbean climate shows no remarkable change during the Holocene in our simulation, which is evidenced by the quasi-persistent AMO feature based on the proxies through the last 8 kyr (Knudsen et al. 2011).

3.2.4 Discussion and Conclusion

The mid-Holocene climatology anomalies from the present-day condition are mainly due to the orbital change. The orbital induced insolation change in the tropical and NH sub-tropical shows the maximum positive anomalies from July to September and the maximum negative anomalies from January to March, which is approximately 1-month lagging or leading with those in the mid-to-high latitudes. The simulated surface air temperature follows this change, resulting in the increased seasonality during the mid-Holocene.

The ENSO activity and its influence on the Caribbean climate demonstrate no considerable change during the mid-Holocene. The climate variability associated with ENSO in the Caribbean is also documented by several proxy records. A lake sediment record from Southern Ecuador indicates that the ENSO has a continuous influence in this region over the last 12,000 years (Moy et al. 2002). They also find that changes of the ENSO frequency become more intensified over the Holocene until 1,200 years ago, which is attributed to the orbital induced change. Clement et al. (2000) conduct a numerical experiment with tropical Pacific driven by orbital forcing and further conclude that, the ENSO is present throughout the Holocene but undergoes a gradual increase from the mid-Holocene to the present, due to the steady warming of the eastern tropical Pacific. Results from the MTM-spectral analysis demonstrate that the periodicity of ENSO in the observation is in a range of 2–8 years, as indicated by studies based on both observation and proxies (e.g. Moy et al. 2002). However, the spectra of the Nino3 indices exhibit quasi-uniformly strong peaks around 3–4 years in CTL and H6 K, without distinguishable change under different background climate conditions (no shown). This unimodal feature of the ENSO periodicity in the simulation might result from the relatively coarse resolution of the ocean model, which is about $5 \times 5°$ along the equator. As a consequence, the Kelvin waves, which signal the start of an ENSO cycle, cannot be well represented. Thus, we argue that a higher resolution should be used in order to simulate a more realistic ENSO character and its possible change during the Holocene.

Previous studies have suggested that the AMO is highly linked to change in the Atlantic Meridional Overturning Circulation (AMOC). In H6 K, due to the insolation change, the AMOC is reduced by more than 2 Sv (Wei and Lohmann 2012). However, the climate influence of the AMO on the Caribbean climate is similar in CTL and H6 K, showing a warming and wetter condition during a positive AMO. The results further supports that the AMO is a quasi-persistent internal ocean–atmosphere variability during large parts of the Holocene (Knudsen et al. 2011).

Previous modelling study (Angeles et al. 2007) suggests that there will be a future warming of approximately 1 °C in SST along with an increase in the rain production during the rainy seasons in the middle of this century. However, the SST seasonality will keep the same variation as present-day situation. It indicates that change in orbital forcing and the increased anthropogenic CO_2 can generate a different climate condition in the Caribbean, although the magnitude of such changes is comparable. Concerning influences of the internal variability on the Caribbean climate in the future, not much effort has been made. Thus, our results can supply useful information to better constrain the prediction of such influences.

3.3 Oceanic $\delta^{18}O$ Variation and its Relation to Salinity in the MPI-OM Ocean Model

Xu Xu[1] ($\boxtimes$), **Martin Werner**[1], **Martin Butzin**[2] **and Gerrit Lohmann**[1]

[1]Alfred Wegener Institute for Polar and Marine Research, Bremerhaven, Germany
e-mail: xu.xu@awi.de
[2]MARUM–Center for Marine Environmental Sciences, University of Bremen, Germany

Abstract Stable water isotope $H_2^{18}O$ is incorporated as into an oceanic general circulation model. A control simulation under present-day climate conditions shows an oceanic $\delta^{18}O$ distribution consistent with available observations, a pattern of enriched water (0.6–2 %) in the subtropics and depleted (−1.8–0 %) at mid-to-high latitudes. The model is also able to capture a quasi-linear relationship between $\delta^{18}O$ and salinity of surface water masses as seen in the observational data. The development of the diagnostic tool in ocean circulation models enables the interpretation of changes in water mass characteristics and marine proxy data during the Earth history.

Keywords Stable water isotopes · Ocean · Hydrological cycle

3.3.1 Introduction

The isotopic composition in marine archives is important to describe the conditions of ancient oceans. Analyses of stable oxygen isotopes from biogenic carbonates are crucial in paleoceanography to indicate paleo-temperature (Broecker 1986), paleo-salinity (Duplessy et al. 1991), and sea level variations (Sower and Bender 1995). These reconstructions are rely on the estimation from the oxygen isotopic composition in sea water, which is closely coupled to the Earth's hydrological cycle, comparable but not identical to salinity (Craig and Gordon 1965).

General circulation models (GCMs) including explicit stable water isotope diagnostics have been used to investigate the relationship between $\delta^{18}O$ values in water and various climate variables. $H_2^{18}O$ has been incorporated in atmospheric (e.g. Joussaume et al. 1984; Hoffmann et al. 1998), oceanic (Schmidt 1998; Wadley et al. 2002), ice sheet (Sima 2006) models, as well as in coupled atmosphere–ocean models (Schmidt et al. 2007). Isotope-enabled ocean models have the potential of applying $\delta^{18}O$ to characterize different water masses, and to understand the spatial and temporal isotopic variations for a quantitative interpretation of their relationship to climate changes.

Here, we show first results with a new isotope-enhanced version of the ocean general circulation model MPI-OM (Marsland et al. 2003) and compare the model output with observed present-day global oceanic distribution of $\delta^{18}O$. This comparison examines the model performance in capturing the main characteristics of the present-day oceanic oxygen isotopic distribution as well as of the $\delta^{18}O$ - salinity relationship in different ocean basins.

3.3.2 Method

3.3.2.1 Ocean Model

The model employed in this study is the Max-Planck-Institute Ocean Model MPI-OM which includes a dynamic-thermodynamic sea ice model (Marsland et al. 2003). The model is implemented on a bipolar orthogonal spherical coordinate system, with the poles located over Antarctic and Greenland, respectively. The second location avoids singularity in the Arctic Ocean while the horizontal resolution is high in the deep-water formation regions of the Arctic Ocean and the northern North Atlantic. Initial conditions for marine temperature, salinity and restoring of sea surface salinity are interpolated from climatology with improved Arctic Ocean information (Steele et al. 2001).

3.3.2.2 Isotope Tracer $H_2^{18}O$

The stable water isotope $H_2^{18}O$ is incorporated as a conserved passive tracer in MPI-OM. The isotopic variations of ocean surface waters are mainly due to the isotopic fractionation occurring during evaporation as well as the isotopic composition of precipitation entering the ocean. The atmospheric forcing is derived from a present-day control simulation as obtained by an atmospheric GCM (Roeckner and Arpe 1995). The atmospheric forcing fields are bilinear interpolated to the MPIOM grid set. The same atmospheric model was enabled with water isotope diagnostics (Werner and Heimann 2002). The daily isotopic content of precipitation and evaporated water vapor over ocean are taken from the same simulation as for the freshwater, heat and momentum fluxes. This ensures maximum consistency between the prescribed climatological forcing and related isotope fluxes.

The isotopic composition of river runoff is calculated from a simple bucket model which drains the continental freshwater fluxes following the topographic slope. Storage effects of lakes and continental ice are neglected to avoid an unbalanced global water cycle in the ocean model.

For the phase transition of water occurring during the formation and melting of sea ice, no change of $\delta^{18}O$ is assumed in our MPIOM model setup, as the oxygen isotopic fractionation between liquid water and ice is very small (Craig and Gordon 1965). This effect provides a difference to salinity where brine release from sea ice provides an important source of high-salinities at high latitudes.

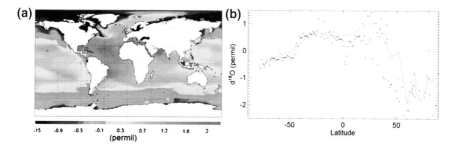

Fig. 3.6 a The annual mean $\delta^{18}O$ distribution at the sea surface as simulated by MPI-OM. **b** The global zonal–mean oxygen isotopic composition of sea surface waters from MPI-OM model simulation (*black line*) and the zonal-mean observed $\delta^{18}O$ values (*red dots*)

As initial condition of $H_2^{18}O$ in the ocean, we set all values to a present-day $\delta^{18}O$ value of 0 % with reference to the ratio of Vienna Standard Mean Ocean Water (Baertsch 1976). No surface restoring of $H_2^{18}O$ is applied.

3.3.2.3 Observation Database of $\delta^{18}O$

The Global Seawater Oxygen-18 Database (Schmidt et al. 1999) contains a collection of over 26,000 observations since about 1950, which offers a unique opportunity to compare the observed and modeled $\delta^{18}O$ values. All data entries of the upper 5 m, where both salinity and $\delta^{18}O$ observations exist, are taken as representative ocean surface water $\delta^{18}O$ values for the comparison with our model results. These observations have been averaged over 0–5 m depth interval and different sample years.

3.3.3 Results

In our performed present-day climate control simulation, the model is run for 3000 years to reach quasi-steady state. All reported results refer to the mean state of the last 100 simulation years.

As illustrate in Fig. 3.6a, the simulated $\delta^{18}O$ depleted regions are mostly located at high latitude oceans where the precipitation is dominant. In sub-tropical ocean regions, the excess in evaporation enriches the isotopic composition of surface waters. Extremely depleted values are seen in marginal see such as the Baltic Sea (-10 to -13 %) and estuaries of Arctic rivers (-6 to -4 %). The most enriched $\delta^{18}O$ values are found in the Red Sea (2.7 %), where evaporation is practically the dominating process that changes the $\delta^{18}O$ of seawater. Because substantial parts of the moisture evaporated over the Atlantic Ocean precipitate over land surfaces and/or the Pacific basin due to the prevalent atmospheric circulation, the Atlantic Ocean between 30°S and 40°N has much higher isotopic

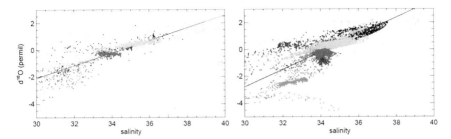

Fig. 3.7 a Relationship between δ^{18}O and salinity S for observed surface water. **b** Relationship between δ^{18}O and salinity S for simulated surface water. *Black* Atlantic Ocean; *Blue* Arctic Ocean; *Green* Labrador Sea; *Yellow* Pacific and Indian Oceans; *Red* Antarctic Ocean

values (δ^{18}O > 1 %) as compared to the Pacific Ocean, and the North Atlantic current transports these enriched water to the Greenland and the Barents Sea. The zonal mean δ^{18}O values exhibit relatively weak variability in the Southern Ocean, which has less enriched subtropics and less depleted mid-to-high latitudes compared to the northern hemisphere (Fig. 3.6b). The most depleted regions are located around 60° and 70°N.

The measurements are averaged along the latitude with 1° interval and compared with the simulated zonal-mean distribution. Although the MPIOM model does not exactly reproduce the observed isotopic values, the simulated spatial structures and basin features are similar to the observations (Fig. 3.6b). These zonal-mean observed data scatter around the simulation results, and the simulated southern oceans is in good agreement with observations. Main model deficits for the general distribution are around 40–50°N, where the model results is approximately 1 % higher. Further analyses of the correspondence between simulated and observed fields at surface suggest a well agreement between these two fields, where the correlation coefficient is 0.78 and the normalized root mean square error is 14.8 %.

Because similar key physical processes (precipitation, evaporation, and runoff) affect the salinity and the isotopic composition of ocean waters, a positive salinity-δ^{18}O relationship can be expected, particularly for ocean surface waters. In our analyses, we have excluded observations and model results with salinity values less than 30 psu to avoid a bias in the comparison towards the areas very strongly affected by river runoff. The model is able to represent the basins features, as most of the model results are have similar salinity-δ^{18}O features in different basin. There is also a nearly horizontal relationship detected in Labrador Sea from the simulation but the observation, where the salinity varies with almost no change at δ^{18}O. The slope of global mixing line from measured data is flatter (0.47) in compare with the model simulation (0.64). The $\delta-$S relationships in tropical and extratropical oceans (Fig. 3.7) are well simulated for the Atlantic and Pacific, where the slope and the δ^{18}O$-$S slopes and end members are consistent with the observations in both regions (not shown).

3.3.4 Discussion and Conclusion

We have included tracer routines into the MPIOM ocean circulation model to simulate a global $\delta^{18}O$ distribution and its basin-specific relationship with salinity under present-day climate conditions. Model limitations, which are mainly related to the relatively coarse spatial resolution of our simulation setup, are especially important for the Arctic estuarine areas and marginal seas. At those regions, our model is unable to reproduce the runoff effects and results in higher $\delta^{18}O$ values. In addition, the observations at high latitudes are mainly measured at summer season, which may induce the absence of sea ice formation influence in the observational data. Therefore there is a nearly horizontal salinity-$\delta^{18}O$ relationship obtained from model simulation in Labrador Sea but the measurements. This relationship indicate the seasonal variation of sea ice in Labrador Sea which increase the salinity due to brine reject during ice formation and lower it when ice melting, but no change on $\delta^{18}O$ because of the neglected fractionation on ice formation and melting in our model.

In general, our model is able to capture the characteristic oxygen-isotopic signatures of water masses found in observations. Despite some regional deficiencies, a good general agreement is found concerning the meridional gradients and basin-specific features of surface ocean waters. The simulated $\delta^{18}O$-salinity relationships of tropical and extra-tropical Atlantic and Pacific regions are also in good agreement with the observations, the $\delta^{18}O$-salinity slope increases with latitude and there exist two clearly different $\delta^{18}O$-salinity relationships in the Atlantic versus Pacific tropical oceans, but almost identical ones in the extra-tropical basins.

As a logical next step, our model will be applied to different paleo-climate conditions to improve our understanding of observed past marine $\delta^{18}O$ changes and its use for paleoceanographic reconstructions.

3.4 Ocean Adjustment to High-Latitude Density Perturbations

Sagar Bora (✉), **Sergey Danilov** and **Gerrit Lohmann**

Alfred Wegener Institute for Polar and Marine research, Bremerhaven, Germany
e-mail: sagar.bora@awi.de

Abstract We examine the influence of mesh resolution on ocean adjustment to high-latitude forcing in a series of numerical simulations performed with a reduced-gravity finite-element model. The mesh resolution is refined down to 5 km at coasts and 20 km at the equator to resolve Kelvin waves. Our numerical experiments show that frequency of forcing in the high latitudes is one of the major factors influencing the amplitude of the signal reaching the equatorial regions.

High frequencies are filtered out through interference, whereas the ocean adjustment at low frequencies is dominated by the large-scale patterns of westward propagating Rossby waves. The shape of Kelvin wave broadens with decreasing resolution for a reasonable range of lateral viscosity, however, the wave speed remains constant for different resolutions. It is argued that future Earth system models may benefit from a high-resolution along the coasts and equator as it warrants better representation of adjustment of the large-scale circulation through wave processes to high-latitude forcing.

Keywords Kelvin waves · Rossby waves · Rossby radius of deformation · Unstructured mesh · Reduced gravity

3.4.1 Introduction

North Atlantic deep water formation (NADW) in the Greenland, Iceland, Norwegian and Labrador Sea drives the large scale ocean circulation, leading to a strong northward heat transport. This heat transport makes North Atlantic about 4 K warmer than North Pacific. Variations in the NADW have been found in paleoclimate records, and it has been suggested that some past climate shifts have been caused by these variations (Dansgaard et al. 1993). The '8200 year BP' cooling event, recorded in the North Atlantic region is another example of abrupt climate change. It has been suggested that the shutdown in the deepwater formation at the North Atlantic due to freshwater input caused by drainage of the Laurentide lakes caused this dramatic regional cooling (Barber et al. 1999; Lohmann 2003). Manabe and Stouffer (1993) showed for the North Atlantic that there is a threshold value between two and four times the preindustrial CO_2 concentration, after which, the thermohaline circulation ceases completely.

Wave processes represent one particular way of transferring these perturbations from one part of the ocean to the other. A reduced gravity set up is used here to answer questions related to the role of mesh resolution, and the sensitivity of the wave signal to the frequency of perturbations. In the framework of a reduced gravity model, the elevation of sea surface height (SSH) represents isopycnal displacement at the thermocline depth in the ocean.

Rotating fluids adjust their pressure and velocity in order to reach a geostrophic balance. After it is reached, the flow is along the isobars. If there is a boundary across the isobars, further adjustment is needed, as no flow across the boundary is possible. This adjustment takes place through Kelvin waves (Gill 1982). The amplitude of these waves is significant only within a distance of the order of the Rossby radius from the boundary. They travel along the coast in only one direction, with the coast on the right side in the Northern Hemisphere and on the left side in the Southern Hemisphere. Since the amplitude of Kelvin waves decreases exponentially from the coast, modelling them requires fine resolution and can be expensive for traditional models. The model used here, is a barotropic shallow-water model (Maßmann et al. 2008) derived from the Finite Element Ocean Model (FEOM), developed at the Alfred Wegener Institute (see, e. g. Wang et al. (2008). Its discretization is based on unstructured triangular surface meshes. Their variable resolution helps us resolve the localized Kelvin waves.

A sinusoidal perturbation in time is excited at the Labrador Sea and travels along the western coast towards the equator as a Kelvin wave. Upon reaching the equator, the wave propagates eastwards along the equator till it reaches the eastern coast. Then, the signal splits up, and it propagates northwards and southwards towards the poles along the coast as a modified Kelvin wave. The interior of the basin is adjusted by westward propagating Rossby waves.

In the following sections, we compare the coastal Kelvin wave propagation on meshes with various resolutions, and also illustrate the role of perturbation frequency in influencing the amplitude of signal reaching the equatorial regions.

3.4.2 Model Set-up

In the first set of experiments performed here, the Atlantic Ocean is represented with a box setup. Figure 3.8a shows the mesh with the highest coastal resolution (7–8 km). The experiments were carried out on 5 meshes with various coastal resolutions, ranging from 7 to 150 km. This was done to compare the propagation of Kelvin waves as a function of resolution. The perturbation is generated by relaxing SSH to a prescribed Gaussian distribution of $3°$ in width centered at 57.5 N and 57.5 W which varies periodically in time with a period of 10 years on all the meshes.

To test the sensitivity of the wave propagation to the frequency of forcing, we used the mesh of the North Atlantic Ocean shown in Fig. 3.8b. We are only interested in the North Atlantic and perform experiments for perturbations with a period lower than a decade. Johnson and Marshall (2002) pointed out that due to the equatorial buffer, effects of high frequency waves are limited to the hemisphere where they originated. The shape of perturbation is as in the previous case and periods of 2, 5, and 10 years are used.

The reduced-gravity model used in all experiments mimics the dynamics of the thermocline layer resting on abyss. Reduced gravity models assume an ocean with homogenous density and a step function in density representing the thermocline. The fluid above the thermocline has higher density than the fluid below, which again moves much slower than that above the thermocline. So, a reduced gravity model assumes that a lighter active layer of fluid sits on a heavier stagnant layer of fluid. The reduced gravity was selected such that the model simulates the first baroclinic mode of Kelvin and Rossby waves. Under a reduced gravity set-up, the SSH perturbations are thought of as representing isopycnal displacement of a stratified ocean.

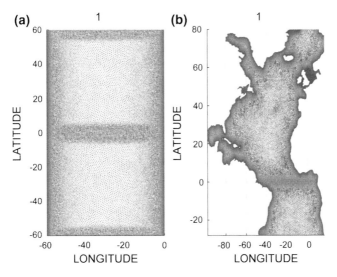

Fig. 3.8 Unstructured mesh representation of the finite element grid for **a** a box model of the Atlantic Ocean and **b** the Atlantic Ocean domain. The *grid cells* vary from 7 km near coast to 150 km in the open ocean

3.4.3 Results

The adjustment of a SSH perturbation in the box setup (Fig. 3.9a) after switching on the forcing, and the propagation of a perturbation on the realistic Atlantic Ocean is shown in Fig. 3.9b. The patterns are similar in both cases. Initially, a Kelvin wave propagates southwards along the western boundary to the equator. At the equator, the coastal Kelvin wave turns eastward, crosses the Atlantic in a couple of months. It splits upon reaching the western boundary and propagates polewards in both hemispheres, whilst shedding westward propagating Rossby waves for interior ocean adjustment.

Figure 3.10a shows how the Kelvin wave on the western coast of the Atlantic Ocean at 40°N decays away from the coast in the box setup. The Rossby radius at 40°N is approximately 40 km. It can clearly be seen that, for the finest resolution of 0.1°, the decay away from the coast is close to the theoretical decay (Gill (1982). As the resolution coarsens, the Kelvin wave broadens. Figure 3.10b shows the phase speed of the Kelvin wave on the western coast of the Atlantic Ocean at 40°N in the box setup, and we can see that the phase speed is independent of the mesh resolution. Figure 3.10c shows the amplitude of the signal reaching the eastern equatorial coast as a function of frequency. It is clear from the figure that the higher the time period, or lower the frequency of the SSH forcing initiating the waves, the higher is the amplitude of the signal crossing the equator and reaching the eastern equatorial coast. The figure compares the amplitude for three different periods, 2, 5, and 10 years.

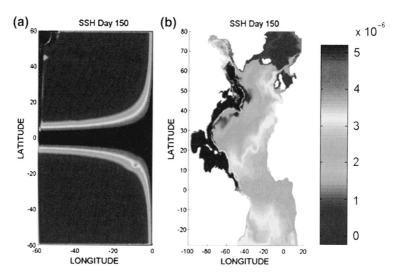

Fig. 3.9 Wave propagation indicated by sea surface height anomalies 150 days after the initiation of the perturbation in **a** the Atlantic box model and **b** the realistic Atlantic Ocean setup

3.4.4 Discussion and Conclusion

The aim of this study was to investigate the dependence of characteristics of Kelvin wave propagation on the mesh resolution, and to analyse the baroclinic response of the North Atlantic with realistic coastlines due to perturbation of isopycnals at various frequencies in the Labrador Sea. After the perturbation has spread as a Kelvin wave along the western coast and across the equator, the signal splits upon reaching the western coast of Africa and propagates polewards. On this stage the Kelvin wave sheds Rossby waves which propagate westward and perform the adjustment of the ocean interior.

Previous studies by Hsieh (1983) have shown that the phase speed of the baroclinic Kelvin wave, in Finite-Difference Numerical Models, are distorted on B-grids, but are more stable on C-grids under varying resolution. However, the off-shore decay structure is better in the B-grid that in the C-grid, as grid scale oscillations occurs in the C-grid. In our experiments, with the finite element reduced-gravity model, we observe that as soon as we resolve the coastal area finer than the local Rossby radius, the off-shore decay structure of Kelvin wave resembles the theoretical one as calculated from the equations in Gill (1982). The phase speed of the Kelvin wave seems to be independent of the resolution.

Our results also show that the amplitude of signal reaching the eastern equatorial coast depends on the frequency of forcing initiating waves. Here we use frequencies of 2, 5 and 10 years, and show that the lower the frequency, the higher is the amplitude of the signal reaching the eastern equatorial coast. Higher the frequency, lower is equatorial response.

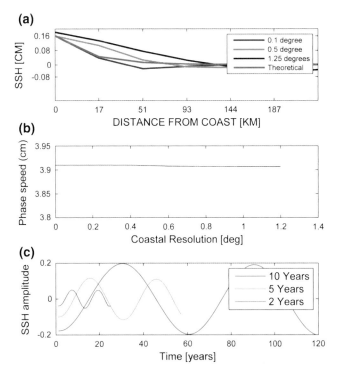

Fig. 3.10 **a** Off-shore decay of the Kelvin wave amplitude for varying resolution. The broadening of the off-shore spread with decreasing resolution can be clearly seen. **b** Phase speed of Kelvin wave as a function of varying resolution. **c** Amplitude of Kelvin wave reaching the equator for perturbation of different frequencies

Our experiments yield information on how the initial Kelvin wave processes are important for density perturbation at the high latitudes as observed by Dickson et al. (1996) during the 1970s. As a logical next step, we examine next the mechanisms as identified here in a coastally-resolving climate model, representing proper stratification of a real ocean, to see the interaction of the waves with a background flow. It is also interesting to see how air-sea interactions influence these processes by the use of a coupled atmosphere–ocean model.

References

Angeles ME, Gonzalez JE, Erickson DJ III, Hernández JL (2007) Predictions of future climate change in the caribbean region using global general circulation models. International J Climatol 27:555–569

Baertschi P (1976) Absolute ^{18}O of standard mean ocean water. Earth Planet Sci 31:341–344

Barber DC, Dyke A, Hillaire-Marcel C, Jennings AE, Andrews JT, Kerwin MW, Bilodeau G, McNeely R, Southon J, Morehead MD, Gagnonk J-M (1999) Forcing of the cold event of 8,200 years ago by catastrophic drainage of Laurentide lakes. Nature 400:344–348

Berger AL (1978) Long-term variations of daily insolation and quaternary climatic changes. J Atmos Sci 35:2362–2367

Braconnot P, Otto-Bliesner B, Harrison S, Joussaume S, Peterchmitt JY, Abe-Ouchi A, Crucifix M, Driesschaert E, Fichefet T, Hewitt C (2007a) Results of PMIP2 coupled simulations of the mid-holocene and last glacial maximum-part 1: experiments and large-scale features. Clim Past 3:261–277

Braconnot P et al (2007b) Results of PMIP2 coupled simulations of the mid-holocene and last glacial maximum—part 2: feedbacks with emphasis on the location of the ITCZ and mid- and high latitudes heat budget. Clim Past 3:279–296

Broecker WS (1986) Oxygen isotope constraints on surface ocean temperatures. Quat Res 26:121–134

Brovkin V, Raddatz T, Reick CH, Claussen M, Gayler V (2009) Global biogeophysical interactions between forest and climate. Geophys Res Lett 36. doi:10.1029/2009GL037543

Clement AC, Seager R, Cane MA (2000) Suppression of El nino during the mid-holocene by changes in the Earth's orbit. Paleoceanogr 15:731–737

Craig H, Gordon LI (1965) Deuterium and oxygen 18 variations in the ocean and marine atmosphere. In: Tongiogi E (ed) Proceeing stable isotopes in oceanographic studies and paleotemperatures, Spoleto, Italy, pp 9–130

Crowley TJ, Kim K-Y (1994) Milankovitch forcing of the last interglacial sea level. Science 265:1566–1568

Crucifix M, Braconnot P, Harrison SP, Otto-Bliesner B (2005) Second phase of paleoclimate modelling intercomparison project. Eos Trans AGU 86

Dansgaard W, Johnsen S, Clausen HB, Dahl-Jensen D, Gundestrup NS, Hammer CU, Hvidberg CS, Steffensen JP, Sveinbjoernsdottir AE, Jouzel J, Bond G (1993) Evidence for general instability of past climate from a 250-kyr ice-core record. Nature 364:218–220. doi:10.1038/364218a0

Dickson RR, Lazier J, Meincke J, Rhines P, Swift J (1996) Long-term co-ordinated changes in the convective activity of the North Atlantic Prog Oceanogr 38(3):241–295(55)

Drysdale RN, Hellstrom JC, Zanchetta G, Fallick AE, Sánchez Gõni MF, Couchoud I, McDonald J, Maas R, Lohmann G, Isola I (2009) Evidence for obliquity forcing of glacial termination II. Science 325:1527–1531. doi: 10.1126/science.1170371

Duplessy JC, Labeyrie LD, Juillet-Leclerc A, Maitre F, Duprat J, Sarnthein M (1991) Surface salinity reconstruction of the North Atlantic ocean during the last glacial maximum. Oceanol Acta 14:311–324

Felis T, Lohmann G, Kuhnert H, Lorenz SJ, Scholz D, Pätzold J, Al Rousan SA, Al-Moghrabi SM (2004) Increased seasonality in Middle East temperatures during the last interglacial period. Nature 429:164–168

Giannini A, Kushnir Y, Cane MA (2000) Interannual variability of caribbean rainfall, ENSO, and the Atlantic ocean. J Clim 13:297–311

Gill A (1982) Atmosphere-Ocean dynamics. Int Geophys Ser, Vol 30. Academic Press, London

Giry C, Felis T, Kölling M, Scholz D, Wei W, Scheffers S (2012) Mid- to late holocene changes in tropical Atlantic temperature seasonality and interannual to multidecadal variability documented in southern caribbean coral records. Earth Plan Sci Lett (in press)

Hagemann S, Gates L (2003) Improving a subgrid runoff parameterization scheme for climate models by the use of high resolution data derived from satellite observations. Clim Dyn 21:349–359

Haug GH, Hughen KA, Sigman DM, Peterson LC, Röhl U (2001) Southward migration of the intertropical convergence zone through the holocene. Science 293:1304

Herold M, Lohmann G (2009) Eemian tropical and subtropical African moisture transport: an isotope modelling study. Clim Dyn 33:1075–1088. doi:10.1007/s00382-008-0515-2

Hibler WD III (1979) A dynamic thermodynamic sea ice model. J Phys Ocean 9:815–846

Hoffmann G, Werner M, Heimann M (1998) Water isotope module of the ECHAM atmospheric general circulation model: a study on timescales from days to several years. J Geophys Res, 103, 14, 16,871–16,896

Hseih W, Davey M, Wajsowicz R (1983) The free Kelvin wave in finite difference models. J Phys Oceanogr 13(8):1383–1397. doi:10.1175/1520-0485

Jansen E et al (2007) Palaeoclimate. In: Solomon S, Qin D, Manning M, Chen Z, Marquis M, Averyt KB, Tignor M, Miller HL (eds) Climate change 2007: the physical science basis. Contribution of working group i to the fourth assessment report of the intergovernmental panel on climate change. Cambridge University Press, Cambridge

Johnson HL, Marshall DP (2002) A theory for the surface Atlantic response to thermohaline variability. J Phys Oceanogr 32(4):1121–1132

Joussaume J, Sadourny R, Jouzel J (1984) A general circulation model of water isotope cycles in the atmosphere. Nature 311:24–29

Jungclaus JH, Keenlyside N, Botzet M, Haak H, Luo JJ, Latif M, Marotzke J, Mikolajewicz U, Roeckner E (2006) Ocean circulation and tropical variability in the coupled model ECHAM5/MPI-OM. J Climate 19:3952–3972

Kaspar F, Cubasch U (2007) Simulations of the eemian interglacial and the subsequent glacial inception with a coupled ocean atmosphere general circulation model. In: Sirocko F, Litt T, Claussen M, Sánchez-Goñi MF (eds) The climate of past interglacials, pp 499–515, Elsevier, Developments in quaternary sciences 7, Chapter 33

Knudsen MF, Seidenkrantz MS, Jacobsen BH, Kuijpers A (2011) Tracking the Atlantic multidecadal oscillation through the last 8,000 years. Nature Com 2:178

Kutzbach JE, Gallimore RG, Guetter PJ (1991) Sensitivity experiments on the effect of orbitally-caused insolation changes on the interglacial climate of high northern latitudes. Quatern Int 10–12:223–229

Lohmann G (2003) Atmospheric and oceanic freshwater transport during weak Atlantic overturning circulation. Tellus 55(A):438–449

Lorenz SJ, Lohmann G (2004) Acceleration technique for milankovitch type forcing in a coupled atmosphere-ocean circulation model: method and application for the holocene. Clim Dyn 23:727–743

Manabe S, Stouffer RJ (1993) Century-scale effects of increased atmospheric CO_2 on the ocean-atmosphere system. Nature 364:215–218

Marsland SJ, Haak H, Jungclaus JH, Latif M, Roeske F (2003a) The Max Planck Institute global ocean/sea-ice model with orthogonal curvilinear coordinates. Ocean Modell 5:91–127

Maßmann S, Androsov A, Danilov S (2008) Intercomparison between finite element and finite volume approaches to model North sea tides. Cont Shelf Res 30(6):680–691. doi:10.1016/j.csr.2009.07.004

Mearns LO, Hulme M, Carter TR, Leemans R, Lal M, Whetton P (2001) Climate scenario development. In: Houghton JT et al. (eds) Climate change 2001: the scientific basis. Cambridge University Press, New York, pp 739–768

Montoya M, von Storch H, Crowley TJ (2000) Climate simulation for 125 kyr BP with a coupled ocean-atmosphere general circulation model. J Clim 13:1057–1071

Moy CM, Seltzer GO, Rodbell DT, Anderson DM (2002) Variability of El nino/Southern oscillation activity at millennial timescales during the holocene epoch. Nature 420:162–165

Otto-Bliesner BL, Marshall SJ, Overpeck JT, Miller GH, Hu A (2006) CAPE Last interglacial project members: simulating arctic climate warmth and icefield retreat in the last interglaciation. Science 311:1751–1753. doi:10.1126/science.1120808

Raddatz TJ, Reick CH, Knorr W, Kattge J, Roeckner E, Schnur R, Schnitzler KG, Wetzel P, Jungclaus J (2007) Will the tropical land biosphere dominate the climate-carbon cycle feedback during the twenty-first century? Clim Dynam 29:565–574

Roeckner E, Arpe K (1995) AMIP experiments with the new max planck institute for meteorology model ECHAM4. In: Proceedings of the "AMIP scientific conference", May 15–19, Monterey, USA, WCRP-Report No. 92, 307–312, WMO/TD-No. 732

Roeckner E, Bäuml G, Bonaventura L, Brokopf R, Esch M, Giorgetta M, Hagemann S, Kirchner I, Kornblueh L, Manzini E, Rhodin A, Schlese U, Schulzweida U, Tompkins A (2003) The atmospheric general circulation model ECHAM5. Part one: model description report No.349. Max Planck Institute for meteorology

Schmidt GA (1998) Oxygen-18 variations in a global ocean model. Geophys Res Lett 25:1201–1204

Schmidt GA, Bigg GR, Rohling EJ (1999) Global seawater oxygen-18 database, http://data.gissgnasa.gov/o18data/, NASA Goddard Institute of Space Science, NY

Schmidt GA, LeGrande AN, Hoffmann G (2007) Water isotope expressions of intrinsic and forced variability in a coupled ocean-atmosphere model. J Geophys Res 112:D10103

Semtner AJ (1976) A model for the thermodynamic growth of sea ice in numerical investigations of climate. J Phys Oceanogr 6:379–389

Sima A, Paul A, Schulz M, Oerlemans J (2006) Modeling the oxygen-isotopic composition of the North American ice sheet and its effect on the isotopic composition of the ocean during the last glacial cycle. Geophys Res Lett 33:L15706

Sowers T, Bender M (1995) Climate records covering the last deglaciation. Science 269:210–214

Steele M, Morley R, Emold W (2001) PHC: a global ocean hydrography with a high-quality arctic ocean. J Climate 14:2079–2087

Sutton RT, Hodson DLR (2005) Atlantic ocean forcing of North American and European summer climate. Science 309:115

Valcke S (2006) OASIS3 user guide (oasis3_prism_2–5). PRISM support initiative report No 3, CERFACS, Toulouse, France, p 64

Valcke S, Caubel A, Declat D, Terray L (2003) OASIS ocean atmosphere sea ice soil users guide. Tech Rep, CERFACS

Wadley MR, Bigg GR, Rohling EJ, Payne AJ (2002) On modeling present-day and last glacial maximum oceanic δ^{18}O distributions. Global Planet Change 32:89–109

Wang Q, Danilov S, Schröter J (2008) Finite element ocean circulation model based on triangular prismatic elements, with application in studying the effect of topography representation. J Geophys Res 113:C05015. doi:10.1029/2007JC004482

Wei W, Lohmann G (2012) Simulated Atlantic Multidecadal Oscillation during the Holocene. J Clim (in press). doi:10.1175/JCLI-D-11-00667.1

Wei W, Lohmann G, Dima M (2012) Distinct modes of internal variability in the Global Meridional Overturning Circulation associated to the Southern Hemisphere westerly winds. J Phys Oceanogr 42:785–801. doi:10.1175/JPO-D-11-038.1

Werner M, Heimann M (2002) Modeling interannual variability of water isotopes in greenland and antarctica. J Geophys Res 107(D1) doi:10.1029/2001JD900253

Chapter 4
Geotectonics

4.1 Continental Deformation of Antarctica During Gondwana's Breakup

Florian Wobbe (✉) and Karsten Gohl

Alfred Wegener Institute for Polar and Marine Research, Germany
e-mail: florian.wobbe@awi.de

Abstract Geophysical data acquired along the Antarctic passive margins constrain the structure and geometry of the deformed continental crust. Crustal thickness estimates range between 7 and 50 km and the Antarctic continent–ocean transition zone (COTZ) extends up to 100–670 km towards the ocean. Continental deformation prior to rifting over a c. 100 million years long time span resulted in crustal stretching factors varying between 1.8 and 5.9. The time span of deformation was sufficiently large and the rifting velocity low enough to extend the margin by up to 300–400 km. Crustal thinning generates a significant subsidence and shallow water passages might already have developed during the rifting phase along the margin. Accounting for accurate continental margin deformation has also consequences for plate-tectonic reconstructions.

Keywords Antarctica · Plate-tectonics · Continental deformation · Crustal thickness · Stretching factor · Plate reconstructions · Continent-ocean transition zone · Magnetic sea floor spreading anomalies

G. Lohmann et al. (eds.), *Earth System Science: Bridging the Gaps between Disciplines*, SpringerBriefs in Earth System Sciences,
DOI: 10.1007/978-3-642-32235-8_4, © The Author(s) 2013

4.1.1 Introduction

Plate-tectonic reconstructions are based on the existence of magnetic anomalies induced by sea floor magnetization parallel to mid-ocean spreading ridges (Cox 1973). Already in 1963 Vine and Matthews proposed that lava, erupted on the sea floor, preserves the polarity of the Earth's magnetic field upon solidification. As the Earth's magnetic field reverses and sea floor spreading along the ridge continues, a set of magnetic stripes with opposite magnetic polarity develops parallel to the spreading ridge (Fig. 4.1a). Anomalies of the same age, so-called isochrons, are identified on both sides of the ridge and fitting these isochrons provides the relative motion of the two diverging plates.

Magnetic sea floor spreading anomalies are only observed in oceanic crust. Magnetic anomalies originating from deformed continental or transitional crust of passive margins cannot be interpreted as spreading anomalies. Therefore, the initial rifting times and extension rates of diverging plates need to be determined by different means. Usually, geological markers, e.g., volcanic material erupted during rift initiation, can be dated and allow estimating the onset of continental rifting. Intra-plate deformation associated with the breakup process prior to the formation of oceanic crust is predominantly located in the continent-ocean transition zone (COTZ). The zone of extended continental and transitional crust is bounded by the landward unstretched continental crust limit (UCCL) and the seaward continent-ocean boundary (COB, Fig. 4.1a). The width of the COTZ and its crustal thickness give indications on the extension rates and the amount of continental extension during the rifting phase (van Wijk and Cloetingh 2002).

Geological samples, gravity data and seismic tomography models suggest that most, if not all, of Antarctica's rifted passive margins consist of extended continental crust (Jokat et al. 2010; König and Jokat 2006; Leinweber and Jokat 2011; Luyendyk et al. 2003; Stagg et al. 2005; Totterdell et al. 2000; Whittaker et al. 2010; Winberry and Anandakrishnan 2004; Gohl 2008; Wobbe et al. 2012). Hence, the region presents an ideal opportunity to study conjugate rifted margins, but this outcome has been hampered by logistic difficulties associated with collecting data adjacent to Antarctica.

For the first time, this study classifies the continental deformation of the circum-Antarctic passive margins based on new data and a review of relevant published data. It also discusses the implications of the margin properties for plate-tectonic and paleo-bathymetric reconstructions.

4.1.2 Method and Data

The motion of a rigid tectonic plate on the Earth's surface may be described by a rotation about a virtual axis through the center of the Earth. Cox and Hart (1986) refer to such rotations as finite rotations and to the intersection of the axis with the

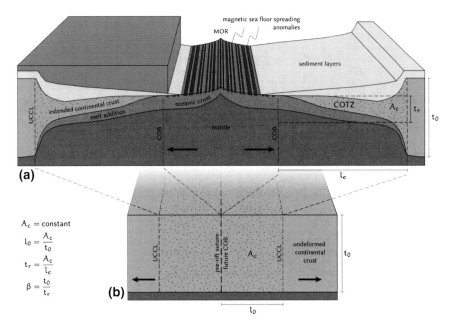

Fig. 4.1 Simplified model of a rifted passive continental margin (**a**) and reconstructed geometry prior to rifting (**b**). *COB* continent–ocean boundary; *COTZ* continent–ocean transition zone; *MOR* mid-ocean spreading ridge; *UCCL* unstretched continental crust limit; β stretching factor; A_c/l_e COTZ area/width; l_0 pre-rift width of COTZ; t_0/t_r initial/extended crustal thickness

Earth's surface as finite pole of rotation. Hence, motion paths of points on a tectonic plate as well as transform faults and flow lines of the generated oceanic crust lie on small circles about this rotation pole. Likewise, the extension during rift initiation is directed parallel to small circles about a finite rotation pole.

If the volume or cross section area, A_c of the deformed transitional crust is constant throughout time and the amount of added material (e.g., melt addition) is known, then a reconstruction of the pre-rift suture is possible (Fig. 4.1b). A_c is determined by integrating the crustal thickness over the width, l_e, of the COTZ along each small circle and the crustal thickness prior to its extension, t_0, is measured at the UCCL. The aforementioned parameters permit the calculation of the pre-rift width of the COTZ, l_0, the mean thickness of the extended crust, t_r, and the stretching factor, β, according to the equations in Fig. 4.1. Both, t_r, and β, are independent from the obliquity of the 2D section with respect to the COTZ, when a three-dimensional continuation of the geological units to either side of the 2D section is presumed (Wobbe et al. 2012).

Crustal thicknesses of the Antarctic passive margins are obtained from teleseismic, seismic and gravity data. Bayer et al. (2009), Reading (2006), and Winberry and Anandakrishnan (2004) estimated crustal thicknesses by using teleseismic earthquakes in Dronning Maud Land, in the Lambert Glacier region, and Marie Byrd Land (Fig. 4.2). Deep crustal seismic data are available in the Weddell Sea,

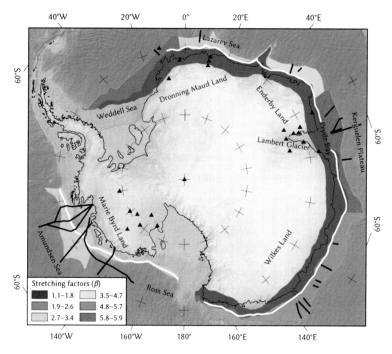

Fig. 4.2 Transitional crust of Antarctica and associated stretching factor (β). *Black profiles and triangles*-location of known crustal thickness; *white line*-reconstructed pre-rift suture. Polar stereographic projection

the Lazarev Sea (Jokat et al. 2004), between the Kerguelen Plateau and Prydz Bay (Gohl et al. 2007a), and in the Amundsen Sea (Gohl et al. 2007b; Wobbe et al. 2012). Further potential field crustal models from Stagg et al. (2005) and Wobbe et al. (2012) off Enderby Land, Wilkes Land, and Marie Byrd Land were used to estimate stretching factors and margin extension.

4.1.3 Results

Crustal thickness estimates of Antarctica's passive continental margins range from 7 to 50 km. Winberry and Anandakrishnan (2004) as well as Wobbe et al. (2012) estimate the continental crust of West Antarctica to be at most 24 km thick, whereas values of less than 50 km are typical in Dronning Maud Land (Jokat et al. 2004; Bayer et al. 2009). In the Lambert Glacier region and off Prydz Bay the crust is thinner than 44 km (Reading 2006; Gohl et al. 2007a). Distal potential field crustal models off Enderby Land and Wilkes Land from Stagg et al. (2005) suggest thicknesses of less than 18 km.

Table 4.1 Rate and lapse of continental deformation of Antarctica's passive continental margins

Domain (conjugate margin)	β-factor	Period in Myr BP (duration)
Weddell Sea (South America)	1.9–2.6	167–147[a] (20)
Dronning Maud Land (Africa)	1.9–3.4	183–154[b] (29)
Enderby Land (India)	3.5–5.9	>118–84[c] (>34)
Wilkes Land (Australia)	4.8–5.9	160–84[d] (76)
Marie Byrd Land (Zealandia)	1.8–3.5[e]	90–(84)62[e] (28)

[a] König and Jokat (2006); [b] Leinweber and Jokat (2011); [c] Jokat et al. (2010); [d] Totterdell et al. (2000); [e] Wobbe et al. (2012)

Stretching factors were derived from the crustal thickness models for the five extensional domains of Antarctica that represent the conjugate to South America, Africa, India, Australia, and Zealandia respectively (Table 4.1). Figure 4.2 illustrates the geographical extent of the transitional crust and the associated stretching factors. This and other recent studies demonstrate that the Antarctic COTZ is generally wider than previously assumed and can extend up to 100–670 km oceanward from the UCCL (Gohl 2008; Jokat et al. 2010; Whittaker et al. 2010; Leinweber and Jokat 2012; Wobbe et al. 2012). The wide COTZ reflects the c.100 million years long timespan of intracontinental deformation that led to the final breakup of Gondwana. The obtained stretching factors roughly correlate with the deformation duration and COTZ width as predicted by numeric lithosphere extension models (van Wijk and Cloetingh 2002).

4.1.4 Discussion and Summary

The separation of South America, Africa, India, Australia and Zealandia from Antarctica was a complex process that stretched over a timespan of c.100 million years. The breakup and rifting of the continents caused intraplate deformation during which the continental crust was stretched by up to 300–400 km with stretching factors of 1.8–5.9. Such large deformation zones develop over prolonged periods (>20 Myr) at low spreading rates (<8 mm/yr), causing the formation of a series of failed rifts that migrate oceanward, referred to as basin migration (van Wijk and Cloetingh 2002). Rifting along Antarctica's margins initiated at different times with varying velocities, generating heterogeneous margin geometries and leading to distinct reconstructions of the pre-rift suture and COTZ. In Marie Byrd Land for instance, pre-rift suture and present-day COB lie as close as 90 km (Wobbe et al. 2012), whereas both are about 400 km apart in Wilkes Land.

The implications of continental deformation on local plate-tectonic reconstructions are substantial: wrong estimation of the pre-rift suture or the neglect of deformation altogether can result in inaccurate reconstructions and large overlaps as illustrated in Fig. 4.3. Apart from the restored plate geometry, the palaeotopography is particularly important for palaeobathymetry models that describe the history of

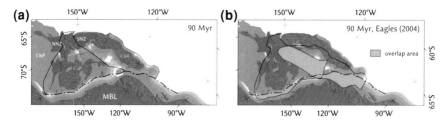

Fig. 4.3 Pre-rift reconstruction models of Marie Byrd Land, Chatham Rise and Campbell plateau considering continental deformation and using finite rotations from Wobbe et al. (2012) (**a**), and model neglecting continental deformation using rotation parameters from Eagles et al. (2004) (**b**). *CaP* Campbell Plateau, *ChP* Challenger Plateau, *ChR* Chatham Rise, *MBL* Marie Byrd Land, *NNZ* North Island of New Zealand, *SNZ* South Island; *black lines*—oceanic plateaus; *dashed line*—pre-rift suture. Base map: crustal thickness of Zealandia and Antarctica, lambert conformal conic projection with central meridian 145°W and standard parallels 72 and 60°S

seafloor topography. Crustal thinning along the Antarctic margins over large periods might have been responsible for subsidence long before the breakup of the continents. Hence, shallow water passages already could have existed during the rifting phase—much earlier than previously assumed.

Acknowledgments Many thanks go to the editors, and an anonymous reviewer, whose comments helped to improve the publication.

References

Bayer B, Geissler WH, Eckstaller A, Jokat W (2009) Seismic imaging of the crust beneath Dronning Maud Land, East Antarctica. Geophys J Int 178(2):860–876. doi:10.1111/j.1365-246X.2009.04196.x

Cox A (1973) Plate tectonics and geomagnetic reversals. W.H. Freeman, San Francisco. ISBN 0-7167-0258-4

Cox A, Hart RB (1986) Plate tectonics: how it works. Blackwell Scientific Publications, Oxford. ISBN 9-78-0-86542-313-8

Eagles G, Gohl K, Larter R (2004) High-resolution animated tectonic reconstruction of the South Pacific and West Antarctic margin. Geochem Geophys Geosyst 5:Q07002. doi:10.1029/2003GC000657

Gohl K (2008) Antarctica's continent–ocean transitions: consequences for tectonic reconstructions. In: Cooper AK, Barrett P, Stagg H, Storey B, Stump E, Wise W, and the 10th ISAES Editorial Team (eds) Proceedings of the 10th international symposium on Antarctic earth sciences, Polar Research Board, National Research Council, U.S. Geological Survey, The National Academies Press, Washington, pp 29–38

Gohl K, Leitchenkov GL, Parsiegla N, Ehlers BM, Kopsch C, Damaske D, Guseva Y, Gandyukhin VV (2007a) Crustal types and continent–ocean boundaries between the Kerguelen Plateau and Prydz Bay, East Antarctica. In: Cooper AK, Raymond CR, and the 10th ISAES Editorial Team (eds) Antarctica: a keystone in a changing world—online proceedings of the 10th ISAES, USGS, USGS open-file report 2007-1047, Short Research Paper 038. doi:10.3133/of2007-1047

Gohl K, Teterin DG, Eagles G, Netzeband G, Grobys J, Parsiegla N, Schlüter P, Leinweber V, Larter R, Uenzelmann-Neben G, Udintsev G (2007b) Geophysical survey reveals tectonic structures in the Amundsen Sea Embayment, West Antarctica. In: Cooper AK, Raymond CR, and the 10th ISAES Editorial Team (eds) Antarctica: a keystone in a changing world—online proceedings of the 10th ISAES, USGS, USGS open-file report 2007–1047, Short Research Paper 047. doi:10.3133/of2007-1047.srp047

Jokat W, Ritzmann O, Reichert C, Hinz K (2004) Deep crustal structure of the continental margin off the explora escarpment and in the Lazarev Sea, East Antarctica. Mar Geophys Res 25:283–304. doi:10.1007/s11001-005-1337-9

Jokat W, Nogi Y, Leinweber V (2010) New aeromagnetic data from the western Enderby Basin and consequences for Antarctic–India break-up. Geophys Res Lett 37(21):L21311. doi:10.1029/2010GL045117

König M, Jokat W (2006) The Mesozoic breakup of the Weddell Sea. J Geophys Res 111:B12102. doi:10.1029/2005JB004035

Leinweber VT, Jokat W (2012) The Jurassic history of the Africa–Antarctica corridor—new constraints from magnetic data on the conjugate continental margins. Tectonophysics 530–531:87–101. doi:10.1016/j.tecto.2011.11.008

Luyendyk BP, Wilson DS, Siddoway CS (2003) Eastern margin of the Ross Sea Rift in western Marie Byrd Land, Antarctica: crustal structure and tectonic development. Geochem Geophys Geosyst 4(10):1090. doi:10.1029/2002GC000462

Reading A (2006) The seismic structure of Precambrian and early Palaeozoic terranes in the Lambert Glacier region, East Antarctica. Earth Planet Sci Lett 244(1–2):44–57. doi:10.1016/j.epsl.2006.01.031

Stagg HMJ, Colwell JB, Direen NG, O'Brien PE, Brown BJ, Bernardel G, Borissova I, Carson L, Close DB (2005) Geological framework of the continental margin in the region of the Australian Antarctic Territory. In: Geoscience Australia Record 2004/25, Petroleum & Marine Division, Geoscience Australia, Canberra

Totterdell J, Blevin J, Struckmeyer H, Bradshaw B, Colwell J, Kennard J (2000) A new sequence framework for the great Australian Bight: starting with a clean slate. APPEA J 40:95–118

Vine FJ, Matthews DH (1963) Magnetic anomalies over oceanic ridges. Nature 199(4897):947–949. doi:10.1038/199947a0

Whittaker JM, Williams S, Kusznir N, Müller RD (2010) Restoring the continent–ocean boundary: constraints from lithospheric stretching grids and tectonic reconstructions. ASEG Ext Abstr 2010(1):1–4. doi:10.1071/ASEG2010ab251

van Wijk J, Cloetingh S (2002) Basin migration caused by slow lithospheric extension. Earth Planet Sci Lett 198(3–4):275–288. doi:10.1016/S0012-821X(02)00560-5

Winberry JP, Anandakrishnan S (2004) Crustal structure of the West Antarctic rift system and Marie Byrd Land hotspot. Geology 32(11):977–980. doi:10.1130/G20768.1

Wobbe F, Gohl K, Chambord A, Sutherland R (2012) Structure and breakup history of the rifted margin of West Antarctica in relation to Cretaceous separation from Zealandia and Bellingshausen plate motion. Geochem Geophys Geosyst , 13, Q04W12, doi: 10.1029/2011GC003742

Chapter 5
Climate Archives

5.1 The Inorganic Carbon System in the Deep Southern Ocean and Glacial-Interglacial Atmospheric CO_2

Franziska Kersten (✉) **and Ralf Tiedemann**

Alfred Wegener Institute for Polar and Marine Research, Bremerhaven, Germany
e-mail: franziska.kersten@awi.de

Abstract After a brief introduction into the marine carbon cycle, the calcite compensation theory and the rain-ratio hypothesis, two theories that may explain glacial to interglacial changes in atmospheric CO_2 concentrations are presented. The validity of these theories in the Southern Ocean is tested with B/Ca-reconstructed carbonate ion concentrations of deep and intermediate waters. Deglacial $[CO_3^{2-}]$ excursions reveal a close relationship between changes in the oceanic inorganic carbon system and atmospheric CO_2, which follow the predictions of the calcite compensation theory on glacial-interglacial timescales. Short-termed $[CO_3^{2-}]$ variations are likely due to the influence of the biological pump and/or changes in circulation patterns.

Keywords Carbon cycle · CO_2 · B/Ca · Carbonate ions · Southern Ocean deep water

5.1.1 Introduction

Ice core records reveal a fluctuation of roughly 100 ppmv in atmospheric CO_2 concentrations from glacials (low CO_2) to interglacials (high CO_2) during pre-industrial times (Fischer et al. 2010). There is no broadly accepted explanation for

G. Lohmann et al. (eds.), *Earth System Science: Bridging the Gaps between Disciplines*, SpringerBriefs in Earth System Sciences,
DOI: 10.1007/978-3-642-32235-8_5, © The Author(s) 2013

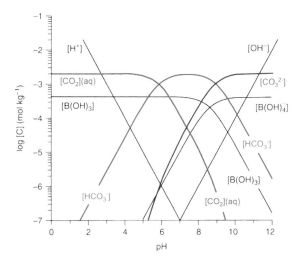

Fig. 5.1 Concentrations of carbonate, borate and water species in seawater versus pH at average salinity (S = 35) and temperature (T = 20 °C). Modified after Emerson and Hedges (2008)

this yet. It has been recognized however, that the deep ocean regulates the carbon exchange between the geosphere, hydrosphere and atmosphere on timescales of thousands of years. Changes in the deep ocean carbon inventory are hence considered to be potential primary drivers of past atmospheric CO_2 variations during glacial/interglacial cycles (Broecker and Peng 1987).

When describing the inorganic carbon system in the seawater, two quantities are of fundamental importance: dissolved inorganic carbon (DIC) and total alkalinity (TA). DIC is defined as the sum of all inorganic carbon species in the water, i.e. DIC = $[CO_2]_{(aq)}$ + $[HCO_3^-]$ + $[CO_3^{2-}]$, where $[CO_2]_{(aq)}$ is the concentration of carbon dioxide in aqueous solution. Alkalinity as a parameter is more complex and can be described as a measure of the charge balance in seawater or to put it more specifically, it is the excess of bases over acids in a mixed electrolyte solution. TA = $[HCO_3^-]$ + $2[CO_3^{2-}]$ + $[B(OH)_4^-]$ + $[H_3SiO_4^-]$ + $[HPO_4^{2-}]$ + $2[PO_4^{3-}]$ + $[OH^-]$ − $[H^+]$ − $[HSO_4^-]$ − $[HF]$ − $[H_3PO_4]$ (Emerson and Hedges 2008).

The three inorganic carbon species introduced above represent roughly 99 % of TA and are thus the most important acid–base pairs contributing to the seawater pH buffering system (Emerson and Hedges 2008). At the present average pH of surface water (pH = 8.2), $[HCO_3^-]:[CO_3^{2-}]:[CO_2]_{(aq)}$ have abundances of roughly 90:9:1 (Fig. 5.1).

Several theories attempt to explain the interaction between inorganic carbon species in the ocean and atmospheric CO_2 on glacial-interglacial timescales. Two of these theories and their key arguments will be discussed here. The *calcite compensation theory* (Broecker and Peng 1987) predicts CO_3^{2-} (carbonate-ion) levels during glacials to be similar to (or lower than) interglacial ones. A conceptual model for deep Pacific $[CO_3^{2-}]$ (Marchitto et al. 2005) demonstrates that excess CO_2 addition to the deep ocean (i.e. removal from the atmosphere) at the onset of glacials results in a pH decrease and thus a deep ocean $[CO_3^{2-}]$ and TA:DIC drop (see Fig. 5.1). Due to the ocean being more acidic, $CaCO_3$ dissolution is enhanced which raises TA:DIC in a 2:1 ratio until the system reaches a new steady state. As stated above, the steady state glacial $[CO_3^{2-}]$ equals the interglacial level provided the fluvial TA input and the global oceanic $CaCO_3$ precipitation rates do not change (Keir 1988).

Then, starting at the Glacial termination, CO_2 is removed from the deep ocean by stratification breakdown and upwelling of deep waters in the Southern Ocean (SO) (e.g. Fischer et al. 2010), and once again added to the atmosphere. This is accompanied by a deep ocean $[CO_3^{2-}]$ excursion, which according to model studies, lasts on the order of thousands of years before returning to a steady state (e.g. Broecker and Peng 1987; Keir 1988). Model results show that $CaCO_3$ compensation alone could explain roughly 30 % of the glacial-interglacial CO_2 fluctuation (Köhler et al. 2005).

On the other hand, the *rain ratio hypothesis* (Archer and Maier-Reimer 1994) states that a change in the ratio between organic and inorganic carbon ($C_{org}:CaCO_3$) which are transported from the surface layer towards depth ('rain') could potentially decouple $CaCO_3$ preservation from $[CO_3^{2-}]$. This is due to the fact that organic matter degradation and $CaCO_3$ dissolution have different effects on DIC and TA. The conceptual model of Marchitto et al. (2005) urges for a glacial increase in $C_{org}:CaCO_3$ that would result in enhanced dissolution of carbonate sediments, raise TA and $[CO_3^{2-}]$ and thus explain the drop in atmospheric CO_2. When deep water $[CO_3^{2-}]$ is high enough to counteract dissolution (at least 50 μmol/kg higher than today), a steady state is reached. At the end of the glacial, the model predicts that the rain ratio would sink again, coeval with CO_2 release.

A key argument of both theories is that glacial-interglacial CO_2 fluctuations are predominantly related to changes in deep-ocean $[CO_3^{2-}]$. Due to the calcite oversaturation of the upper ocean, it can be assumed that $[CO_3^{2-}]$ changes there are of only minor importance (Marchitto et al. 2005).

5.1.2 Objective/Study Area/Materials and Methods

In order to test the theories outlined above, knowledge about past deep ocean CO_3^{2-} concentrations is needed. Marine sediments represent essential archives of past climate conditions and allow reconstructing changes in the carbonate ion signatures of different water masses over time.

This study was carried out in the Southern Ocean (SO), a key area for water mass exchange between all world oceans. The importance of this region with respect to the carbon cycle is demonstrated by a high correlation between Antarctic temperature and atmospheric CO_2 concentrations that is documented for the past 800.000 years (Fischer et al. 2010).

We present results from two cores that were retrieved from the slope of Chatham Rise (East of New Zealand), during Polarstern cruise ANT-XXVI-2 (Fig. 5.2a).

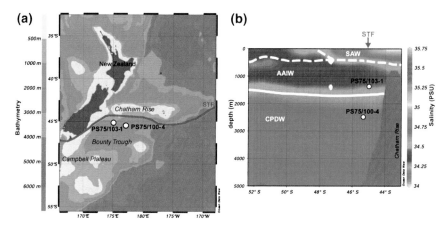

Fig. 5.2 Two views of the study area with position of studied cores: **a** map view with bathymetry and oceanographic features and **b** GLODAP salinity (in practical salinity units) profile along 175°W (Key et al. 2004). The modern position of water masses is sketched. SAW = Subantarctic Water, AAIW = Antarctic Intermediate Water, CPDW = Circumpolar Deep Water. The modern position of the subtropical front (STF), according to Orsi et al. (1995), is indicated in *red*

Cores PS75/100-4 and PS75/103-1 stem from 2498 to 1390 m water depth, respectively. Sediments from the deeper core were bathed by Circumpolar Deep Water (CPDW), while samples from the shallower core are assumed to predominantly reflect the influence of Antarctic Intermediate Water (AAIW) (Fig. 5.2b). In order to reconstruct $[CO_3{}^{2-}]$ changes in the CPDW and AAIW since the Last Glacial Maximum (LGM), benthic foraminifer species *C. wuellerstorfi* and *C.cf. wuellerstorfi* were picked and their B/Ca ratios analysed on a HR-LA-ICP-MS (High Resolution-Laser Ablation-Inductively Coupled Plasma-Mass Spectrometer) at the Geomar Helmholtz Centre for Ocean Research Kiel. Yu and Elderfield (2007) documented a linear relationship between benthic foraminiferal B/Ca and deep water $\Delta[CO_3{}^{2-}]$ (degree of calcite saturation). Following their approach, past deep water $[CO_3{}^{2-}]$ was reconstructed using: $[CO_3{}^{2-}] = \Delta[CO_3{}^{2-}] + [CO_3{}^{2-}]_{sat}$, where $[CO_3{}^{2-}]_{sat}$ is the carbonate ion concentration at saturation.

$[CO_3{}^{2-}]_{sat}$ was calculated with CO_2sys (Pierrot and Wallace 2006). Required input parameters (TA, DIC, [Si], [P], temperature, salinity and pressure) were taken from ANT XXVI-2 cruise-data (Rhee, pers. comm.) and, where not available, estimated from nearby GLODAP sites (Key et al. 2004).

5.1.3 Results/Discussion

Reconstructed deep water $[CO_3{}^{2-}]$ for PS75/100-4 and PS75/103-1 are shown for the past ca. 25 kyrs in comparison to atmospheric CO_2 (Monnin et al. 2006)

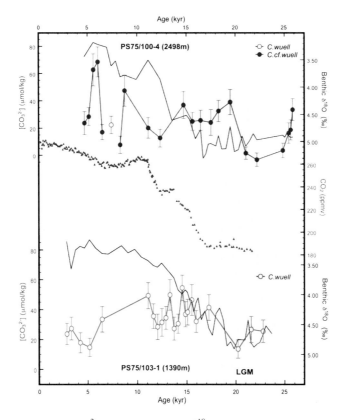

Fig. 5.3 Reconstructed $[CO_3^{2-}]$ and mixed benthic $\delta^{18}O$ for PS75/100-4 and PS75/103-1 for the past ca. 25 kyrs. Error bars contain analytical uncertainty (2σ) and the intercept range of the reconstruction equation (Yu and Elderfield 2007). Atmospheric CO_2 from EPICA DOME C (EDC1 timescale) is given for comparison (Monnin et al. 2006). The approximate interval of the LGM is shaded in *grey*

(Fig. 5.3). During this interval, carbonate ion concentrations range from -3.3 to 68.6 μmol/kg in core PS75/100-4 and between 13.5 and 54.5 μmol/kg in core PS75/103-1. Overall, foraminiferal B/Ca and accordingly reconstructed deep water $[CO_3^{2-}]$ appear to be highly variable on short timescales. Raitzsch et al. (2011) documented a heterogeneous distribution of Boron in foraminiferal shells which might be due to ontogenetic effects. Despite this intra-shell variability, they find a linear correlation between B/Ca and the degree of deep water calcite saturation in agreement with previous studies (e.g. Yu and Elderfield 2007).

Note, that the LGM as indicated by benthic $\delta^{18}O$, appears slightly later in core PS75/100-4 than in PS75/103-1. This might be an effect of age-model uncertainties and can hopefully be resolved with additional $\delta^{18}O$ data. Trends in the data can nevertheless be reliably interpreted.

There are two peaks in the CPDW core between 25 and 19 kyrs BP (before present), which likely reflect internal mechanisms, such as deep-sea $CaCO_3$ dissolution events, seeing that atmospheric CO_2 concentrations stayed constant throughout this time. Both excursions are transient and CO_3^{2-} concentrations decrease towards a steady state that is stable for 4 and 2 kyrs respectively (see Fig. 5.3, periods from ca. 25–21 kyrs BP and between ca. 17.5 and 15.5 kyrs BP). In the shallower core, a drop in $[CO_3^{2-}]$ around 20 kyr BP from relatively constant levels can be observed. This negative excursion appears to be roughly coeval with the $[CO_3^{2-}]$ peak in PS75/100-4, an indication that deep and intermediate water mass histories were decoupled during this time. When the deglacial CO_2 rise begins around 17 kyrs BP, $[CO_3^{2-}]$ in the deeper core remains in a steady state until roughly 15.5 kyrs BP, when it shows an initial increase. Conversely, $[CO_3^{2-}]$ reconstructed from PS75/103-1 rises until roughly 17 kyrs BP, when a transient decrease is observed. The different response time in both cores might be another reflection of temporarily divergent AAIW and CPDW ventilation histories.

It has been argued, that the deep SO was less well ventilated during the Last Glacial (Hodell et al. 2003), effectively impeding exchange between the ocean and the atmosphere and thus reducing atmospheric CO_2 (Köhler et al. 2005). Stable carbon isotope data from Elmore et al. (pers. comm.) reveal, that deep water near New Zealand remained least well ventilated until well after the LGM, whereas ventilation ages of the shallower AAIW and SAW (Subantarctic Water) decrease dramatically at this time. We thus argue that longterm trends in our data can be explained in line with the *calcite compensation theory* (Broecker and Peng 1987), seeing that both cores show deglacial $[CO_3^{2-}]$ peaks. Calcite compensation tends to bring carbonate ion concentrations back to their initial values, which explains the transient nature of the deglacial $[CO_3^{2-}]$ rise. Several negative and positive excursions occur throughout the past ca. 15 kyrs, possibly mirroring changes in the efficiency of the biological pump (Sigman et al. 2010) and/or water mass reorganization. Superimposed on these effects, calcite compensation drives the system towards enhanced $CaCO_3$ dissolution or preservation until $[CO_3^{2-}]$ equal to LGM values are reached again during the Holocene.

5.1.4 *Outlook*

Analysis of core-top samples will provide insights into the most recent evolution of carbonate ion concentrations in AAIW and CPDW, enabling us to directly compare $[CO_3^{2-}]$ reconstructed from foraminiferal B/Ca with in situ measurements. Single species $\delta^{18}O$ measurements are needed to refine the age-model and thus allow a time-sensitive interpretation of deep and intermediate water $[CO_3^{2-}]$ changes relative to atmospheric CO_2 concentrations. Stable carbon isotopes measured on the same samples will be used to further constrain water mass histories in the study area, in order to test whether and during which interval AAIW was decoupled from underlying CPDW.

5.2 The Significance of the Long Lived (>400 Years) Bivalve *Arctica Islandica* as a High-Resolution Bioarchive

Jacqueline Krause-Nehring[1] (✉), **Thomas Brey**[1], **Simon R. Thorrold**[2], **Andreas Klügel**[3], **Gernot Nehrke**[1], **and Bernd Brellochs**[4]

[1]Alfred Wegener Institute for Polar and Marine Research, Bremerhaven, Germany
e-mail: thomas.brey@awi.de
[2]Biology Department MS 50, Woods Hole Oceanographic Institution, Woods Hole, MA, USA
[3]Fachbereich Geowissenschaften, University of Bremen, Bremen, Germany
[4]Emil-von-Behring-Straße 37, D-85375 Neufahrn, Germany

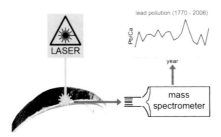

Abstract Information about past environmental conditions is preserved in the elemental signature of biogenic marine carbonates. Thus, trace elements to calcium ratios (Me/Ca) of biogenic calcium carbonates, such as bivalve shells, are often used to reconstruct past environmental conditions at the time of carbonate formation (Foster et al. 2008). In this study, we examine the suitability of the long-lived (>400 years) bivalve *Arctica islandica* as a high-resolution bioarchive by measuring Me/Ca ratios in the shell carbonate. Pb/Ca concentrations in *A. islandica* shells reflect anthropogenic gasoline lead consumption and further provide a centennial record of lead pollution for the collection site off the coast of Virginia, USA. With *A. islandica* shells from the North Sea we test the hypothesis that Ba/Ca and Mn/Ca ratios are indicators of the diatom abundance. Our results indicate that statistically both ratios correlate well with the diatom abundance, and yet, on a year-to-year base, there is no consistent reflection of diatom abundance patterns in the Ba/Ca and Mn/Ca annual profiles. These findings indicate that primary production affects Ba/Ca and Mn/Ca shell ratios, though we suggest that both elements are coupled to primary production through different processes and are affected by further, yet unknown processes.

Keywords *Arctica islandica* · Bivalve · Bioarchive · Biogenic carbonate · Trace elements · Lead · Barium · Manganese · Gasoline lead pollution · Ocean production

5.2.1 Introduction

"Bioarchives" are organisms that grow permanent hard body parts by periodic accretion of biogenic material. These hard parts, e.g., bivalve shells, record the ambient environmental conditions throughout the organism's life-span. In the terrestrial system, trees (dendrochronology) and in the marine environment calcium carbonate parts of corals, bivalves, and finfish are used as such archives (sclerochronology). This section focuses on the long-lived (>400 years) bivalve *Arctica islandica* as a high-resolution bioarchive. In several studies we analyze the biogeochemistry in terms of trace element to calcium ratios (Me/Ca) of *A. islandica* shells to reconstruct environmental parameters of the marine ecosystem over time scales of decades to centuries.

Sample treatment prior to Me/Ca analysis often includes chemical removal of organic matter from the biogenic calcium carbonate (Gaffey and Bronnimann 1993). The efficiency of this approach, however, remains questionable and chemical treatment itself may alter the outcome of subsequent Me/Ca analysis (Love and Woronow 1991). Thus, we first examine the efficiency of eight chemical treatments and their impact on the carbonate composition (for further details see Krause-Nehring et al. 2011a) (**Effect of sample preparation**).

Next, we aim at reconstructing environmental history by measuring trace elements along the growth trajectory of *A. islandica* shells. We determine Pb/Ca ratios in an *A. islandica* shell to examine influxes of lead into the seawater and to establish a centennial record of anthropogenic lead pollution at the collection site off the coast of Virginia, USA (Krause-Nehring et al. 2012) (**Lead as a pollution tracer**). In addition, we measure Ba/Ca and Mn/Ca ratios in three *A. islandica* shells collected off the island of Helgoland and correlate our results with the diatom abundance in the North Sea to evaluate both ratios as potential indicators of ocean primary production (Krause-Nehring et al. 2011b) (**Barium and manganese as indicators of primary production**).

5.2.2 Methods

5.2.2.1 Effect of Sample Preparation

To examine the efficiency and side effects of 8 chemical treatments, we conducted a systematic study on inorganic calcium carbonate and *A. islandica* shell powder. We combined different analytical techniques, such as

(1). inductively coupled plasma-mass spectrometry (ICP-MS),
(2). nitrogen (N) analyses, and
(3). X-ray diffractometry (XRD)

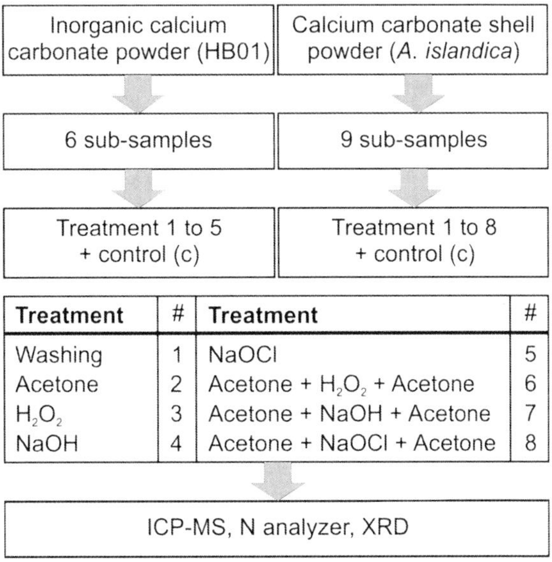

Fig. 5.4 Preparation, treatment (control "c"; treatment 1–8), and subsequent analyses (ICP-MS, N analyzer, XRD) of inorganic (HB01) and organic (*A. islandica*) calcium carbonate powder samples

Inorganic calcium carbonate powder (HB01)	Calcium carbonate shell powder (*A. islandica*)
6 sub-samples	9 sub-samples
Treatment 1 to 5 + control (c)	Treatment 1 to 8 + control (c)

Treatment	#	Treatment	#
Washing	1	NaOCl	5
Acetone	2	Acetone + H_2O_2 + Acetone	6
H_2O_2	3	Acetone + NaOH + Acetone	7
NaOH	4	Acetone + NaOCl + Acetone	8

ICP-MS, N analyzer, XRD

to analyze the impact of each treatment on

(1). Me/Ca ratios,
(2). organic matter content (using N as a proxy), and
(3). the composition of the carbonate and of newly formed phases (Fig. 5.4).

5.2.2.2 Lead, Barium, and Manganese Measurements

Prior to Me/Ca analyses, we embedded each shell in epoxy resin and cut a narrow section along the (red) line of strongest growth (Fig. 5.5a). Next, we ground the section with sandpaper until the annual growth lines were clearly visible (Fig. 5.5b). Finally, we used a laser ablation system connected to an inductively coupled plasma-mass spectrometer (LA-ICP-MS) for element analyses (Pb/Ca, Ba/Ca, Mn/Ca) of the shell carbonate (Fig. 5.5c). In the end, we either assigned each laser spot a specific year using the growth lines as year markers (inter-annual Pb/Ca variations) or converted the location of each laser spot between two adjacent growth lines into a point in time during the year (Ba/Ca and Mn/Ca intra-annual variations).

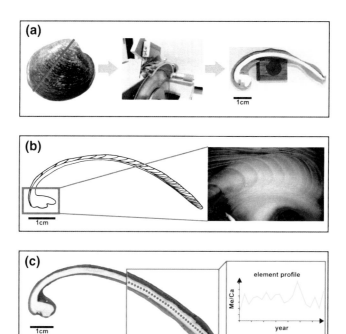

Fig. 5.5 **a** and **b** Preparation of an *A. islandica* shell for subsequent element analyses of the shell carbonate **c** using a laser ablation-inductively coupled plasma-mass spectrometer (LA-ICP-MS). The *red line* in **a** indicates the line of strongest growth

5.2.3 Results and Discussion

5.2.3.1 Effect of Sample Preparation

Our results indicate that the different treatments

(1). alter the Me/Ca ratios (Fig. 5.6),
(2). vary in their ability to remove organic matter (with NaOCl being the most efficient), and
(3). can alter the phase composition of the sample (e.g., $Ca(OH)_2$ formation during treatment 4).

Thus, chemical removal of organic matter prior to Me/Ca analyses has to be applied with extreme caution (for further details see Krause-Nehring et al. 2011a).

Fig. 5.6 Effects of treatments (control "c"; treatments 1–8) on the Me/Ca ratios of the (*left*) HB01 and of the (*right*) *A. islandica* shell powder samples. (*grey*: no significant difference between the treated sample and the control, *red*: significant increase, *green*: significant decrease)

HB01 powder				c	A. islandica shell powder			
Mg/Ca	Sr/Ca	Ba/Ca	Mn/Ca		Mg/Ca	Sr/Ca	Ba/Ca	Mn/Ca
				1				
				2				
				3				
				4				
				5				
				6				
				7				
				8				

Legend: no significance / decrease / increase

5.2.3.2 Lead as a Pollution Tracer

Our results indicate that the lead profiles we obtain from *A. islandica* shells reflect local influxes of lead into the seawater. The Pb/Ca profile we measure between 1770 and 2006 in an *A. islandica* shell collected off the coast of Virginia, USA, is clearly driven by anthropogenic lead emissions due to gasoline lead combustion which are transported eastwards from the North American continent to the Atlantic Ocean by westerly surface winds (Fig. 5.7). Depending on the prevalent sources of lead at certain locations, the lead profiles of *A. islandica* shells may as well be driven by random natural influxes of lead into the water or various other sources of lead (e.g., dumping of sewage sludge or munitions; see Krause-Nehring et al. 2012). Our findings support the applicability of Pb/Ca analyses in *A. islandica* shells to reconstruct anthropogenic lead pollution at specific locations. In addition, we provide a centennial record of lead pollution for the collection site off the coast of Virginia, USA. For comparison of *A. islandica* lead profiles from different boreal sites (Iceland, USA, and Europe) (see Krause-Nehring et al. 2012).

5.2.3.3 Barium and Manganese as Indicators of Primary Production

Over several decades, we find a significant correlation between the Mn/Ca and Ba/Ca ratios of three *A. islandica* specimens collected off the island of Helgoland and the diatom abundance in the North Sea (Krause-Nehring et al. 2011b). Nevertheless, the annual Ba/Ca (summer peak) and Mn/Ca profile (spring and summer peak) do not resemble the annual diatom profile (spring and summer peak) in a consistent manner (Fig. 5.8). Thus, we conclude that primary production does affect Ba/Ca and Mn/Ca shell ratios, though we suggest that both elements are coupled to primary production through different processes. We suggest that peak concentrations of barium in bivalve shells result from sudden fluxes of barite to the sediment water interface as a consequence of phytoplankton blooms (Stecher et al. 1996), and that this mechanism involves an extended time delay (~ 3.5 months) between diatom blooms and Ba/Ca peaks in *A. islandica* shells, as observed in our study (Fig. 5.8: ~ 3.5 months time

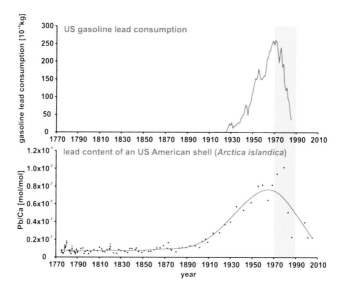

Fig. 5.7 *Top graph*: US gasoline lead consumption (in 10^{+6}kg) [modified after Nriagu (1990)]. *Bottom graph*: Pb/Ca profile (in mol/mol) between 1770 and 2010 determined in an *A. islandica* shell collected off the coast of Virginia, USA, with the *black dots* indicating the annual Pb/Ca ratios (± 1 standard error for years with >1 sample spot) and the *red line* being a cubic spline trendline ($\lambda = 8,000$). The *yellow bar* indicates the time of maximum gasoline lead emissions (1980 $\pm$ 10 years)

lag between the spring bloom and subsequent Ba/Ca summer peak). The second diatom bloom in summer would cause another increase in barite in winter which coincides with the winter growth inhibition (mid-December to mid-February) (Schöne et al. 2005) of *A. islandica*, and is thus, not recorded by the shell. Mn/Ca ratios, on the other hand, seem to be coupled to diatom abundance both through direct influx of manganese to the sediment water interface or through remobilization of manganese from sediments during post-bloom reductive conditions, and thus, instantly record any phytoplankton debris reaching the ocean floor (Krause-Nehring et al. 2011b).

5.2.4 Conclusion

Since environmental data is often limited in time and space, bioarchives provide valuable information to reconstruct past environmental conditions. The bivalve *A. islandica* is an important bioarchive due to its longevity, wide distribution, and long-term occurrence throughout earth history. Our results demonstrate that both long-term and high-resolution records of environmental history can be extracted from *A. islandica* shells. They further illustrate, however, that it is crucial to

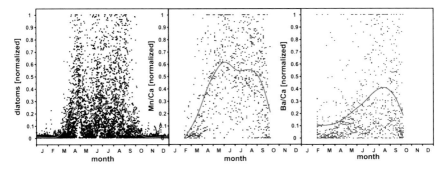

Fig. 5.8 *Left*: Typical annual profile of the diatom abundance measured off the coast of Helgoland as part of the Helgoland Roads timeseries (Wiltshire and Dürselen 2004). All data points are plotted over the course of one calendar year (J = January to D = December) after filtering (removal of the upper 5 % and lower 10 % of the data) and normalization (minimum = 0; maximum = 1) of the data. Superimposed is the corresponding cubic spline trendline ($\lambda = 0.025$). *Centre* and *right*: Typical annual (*centre*) Mn/Ca and (*right*) Ba/Ca profile obtained from three *A. islandica* shells collected off the coast of Helgoland. All data points are plotted over the course of one calendar year (J = January to D = December) after detrending (removal of linear trends, where necessary), filtering (removal of the upper 5 % and lower 10 % of the data), and normalization (minimum = 0; maximum = 1) of the data. Superimposed are the corresponding cubic spline trendlines ($\lambda = 0.0045$)

understand the mechanistic links between bivalve shell chemistry and environmental parameters in order to extract valuable information from bivalve shells. Future studies on the biogeochemistry and growth morphology of *A. islandica* shells will facilitate our understanding of environmental processes within the field of earth system science.

5.3 Sub-Annual Resolution Measurements of Dust Concentration and Size in Different Time Slices of the NorthGRIP Ice Core

Katrin Wolff[1,2] (✉), **Anna Wegner**[1] and **Heinz Miller**[1]

[1]Alfred Wegener Institute for Polar and Marine Research, Bremerhaven, Germany
e-mail: anna.wegner@awi.de
[2]MARUM-Center for Marine Environmental Sciences, Bremen, Germany

Abstract Five ice sections extracted from the NorthGRIP ice core have been analysed in sub-annual resolution in terms of dust concentration and size distribution. The ice samples, covering time slices between 9900 and 35000 yrs before A.D. 2000 (according to Greenland Ice Core Chronology 2005), are taken from the early Holocene, the Allerød Interstadial, the Last Glacial Maximum, the late glacial, and the Dansgaard-Oeschger event 7. The seasonally resolved dust concentration records show systematic variations with one clear dust peak within each year, secondary maxima are clearly visible in the early Holocene and Allerød Interstadial. Pronounced seasonal amplitudes of a factor of ~ 7 to ~ 11 are found in the dust concentration. Seasonal variations of the size distribution were observed with the tendency towards coarser (finer) particles during the dust maxima (minima). Mostly, the dust volume distributions are found to be lognormal; the lognormal mode μ varies on average between 2.03 and 2.22 µm. The largest mode μ was mostly found before or behind the dust maxima.

Keywords Dust · Greenland · NorthGRIP · Seasonality · Holocene · Glacial

5.3.1 Introduction

Mineral dust is an important component of the climate system and plays multiple roles in physical and biogeochemical exchanges between the atmosphere, land surface and ocean (Harrison et al. 2001). The amount, size distribution and composition of dust deposited on polar ice sheets hold valuable information about climate conditions of source areas, long-range transport and deposition processes (Biscaye et al. 1997). The Taklamakan desert is the primary source in the dusty spring season at present day for dust transported to Greenland, whereas the Mongolian Gobi is the source area during the low-dust season from summer to winter (Bory et al. 2002). The concentration of dust found in ice cores is extremely variable over different climate periods, with 10–100 times higher dust concentrations in ice from the last glacial period compared to the Holocene (Steffensen 1997). Higher dust content in Greenland ice cores during glacial times are explained by increased desert area in central Asia and changes in wind strength (Ruth et al. 2003; Mahowald et al. 2006). Sub-annual resolution dust measurements provide detailed information about past abrupt climate fluctuations (Steffensen et al. 2008).

Most work has focused on long-term variations of dust concentration and size distribution in Greenlandic deep ice cores using multi-annual mean values (e.g. Steffensen 1997; Ruth et al. 2003). There are only a few studies dealing with the seasonality of dust concentration in Greenland ice cores, which are only investigating very recent time scales (e.g. Bory et al. 2002).

Here, we present dust concentration and size distribution in sub-annual resolution measured in the NorthGRIP ice core from the early Holocene and for the first time also from the last glacial and glacial/interglacial transition. Furthermore, we investigate the relationship between mass concentration and particle size and finally use the seasonal variations to determine annual layers.

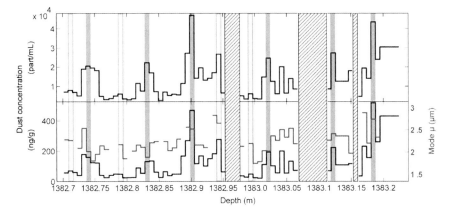

Fig. 5.9 Mass concentration, mode μ (bottom) and number concentration of dust. Gaps in the graph of size distribution arise from lack of distribution fitting. Certain annual layers are indicated with full grey bars, uncertain layers with open bars. Hatched areas represent gaps due to breaks in the ice

5.3.2 Method

In this study we present dust concentration and size distribution measured in a sub-annual resolution on 5 selected ice segments, each 55 cm long, extracted from the NorthGRIP ice core for the early Holocene (~ 9980 year BP), Allerød Interstadial (IS, ~ 13640 year BP), Last Glacial Maximum (LGM, ~ 20870 year BP), late glacial period (~ 25920 year BP) and Dansgaard-Oeschger event 7 (DO7, ~ 35420 year BP). The selected depth resolution for each ice segment ranged between 2 and 7 mm depending on the accumulation rate, corresponding to a temporal resolution of at least 7 samples per year. The amount of particles and the size distribution for discrete samples were measured using a Multisizer 3 Coulter Counter (CC). A lognormal function was fitted to the volume distribution to obtain the lognormal mode μ used as a parameter for the dust size. The fitting procedure was applied to particles in the size range from 1.0 to 7 μm as larger particles have too high uncertainties due to the low counting statistics. The mass concentration was calculated assuming a mean particle density of 2.5 g/cm^3.

The seasonal variations in dust concentration are used to determine annual layers by locating a dust maximum followed by a minimum. Further criteria are a minimum distance of two samples between two maxima, minimal amplitude of a factor ~ 2–3 between maximum and minimum and a minimal peak width of two samples. If all of these criteria are fulfilled the annual layers are defined as certain, otherwise as uncertain.

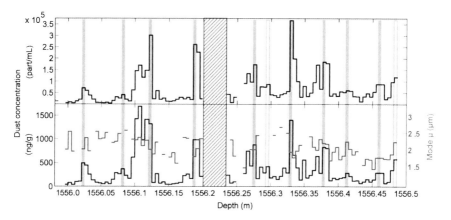

Fig. 5.10 As Fig. 5.9, but for Allerød IS

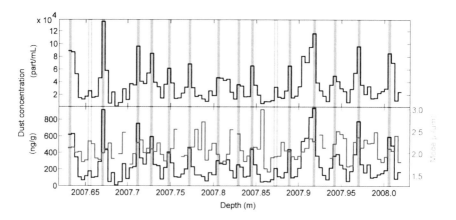

Fig. 5.11 As Fig. 5.9, but for DO 7

5.3.3 Results

Seasonal variations of the mass concentration are observed in all analysed ice intervals. The Holocene ice interval shows the lowest mass concentration of $\sim 120 \pm 50$ µg kg^{-1} with seasonal amplitudes of a factor of $\sim 7.3 \pm 4.7$ (certain annual layers) and $\sim 4.8 \pm 4.0$ (certain and uncertain annual layers) respectively (Fig. 5.9). Often, secondary maxima can be determined in the course of the year. The profile of the mode μ shows numerous gaps mainly during the low-dust periods, when no lognormal distribution could be fitted due to the low counting statistics. The mean of the mode μ was found at 2.22 ± 0.14 µm. There is a tendency towards larger particles during high-dust periods, a moderate correlation

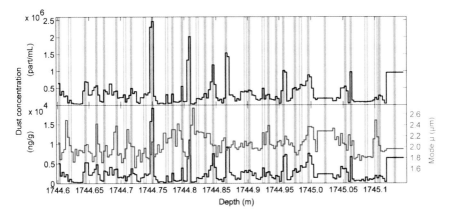

Fig. 5.12 As Fig. 5.9, but for LGM

between concentration and size (r = 0.62 at the 99.5 % significance level) was found.

The Allerød IS ice section is also characterized by clear annual maxima and frequently occurring secondary maxima, which mostly succeed the dust peak (e.g. 1556.27 m, Fig. 5.10). A mean mass concentration of $\sim 330 \pm 190$ µg kg^{-1} was measured. The seasonal amplitudes of the mass concentration vary between $\sim 10.7 \pm 7.3$ (certain years) and 9.1 ± 7.2 (certain and uncertain years, Fig. 5.10). Again, the profile of the mode µ has numerous gaps, especially in the low-dust period. Considering this, the mean of the mode µ in the Allerød IS was found at about 2.07 ($\pm$ 0.20) µm. There is a slight tendency towards larger particles in high-dust periods, however no clear correlation between dust concentration and size was found (r = 0.20 at the 95 % significance level).

The DO 7, the oldest IS analyzed in this study, shows an averaged mass concentration of $\sim 250 \pm 70$ µg kg^{-1} and pronounced seasonal variations with an amplitude of $\sim 7.7 \pm 4.3$ (certain annual layers) and $\sim 7.0 \pm 4.3$ (certain and uncertain annual layers), respectively (Fig. 5.11). The shape of the seasonal cycle differs from the previous intervals; in most cases a moderate increase to the dust maximum is followed by a slow decrease to the low-dust level. Also, secondary maxima are not clearly visible. The mean mode µ was found at $\sim 2.04 \pm 0.08$ µm and shows seasonal variations. The variation of the mode µ and logarithmic mass are not in phase, however, a weak significant correlation of r = 0.39 (at the 99.5 % significance level) was found.

The ice sections belonging to colder periods of the last glacial, the LGM and late glacial period, are characterized by higher dust concentrations of $\sim 2500 \pm 1360$ µg kg^{-1} and $\sim 5600 \pm 3500$ µg kg^{-1}, respectively (Figs. 5.12 and 5.13). The dust concentration in the LGM (late glacial period) shows an increase of a factor of ~ 21 (~ 47) compared to the early Holocene. Seasonal variations with an amplitude of a factor of $\sim 10 \pm 12$ (certain annual layers) and

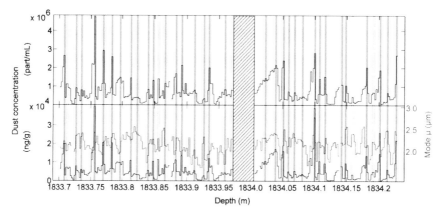

Fig. 5.13 As Fig. 5.9, but for late glacial period

~8–10 ± 11 (including uncertain annual layers) are found. Both periods are characterized by a similar shape of the seasonal cycle with a sharp increase to the dust maxima, followed by a moderate decrease to the low-dust level later on during the year. On average, slightly smaller particles were observed in the LGM (~2.03 ± 0.22 μm) than in the late glacial (~2.15 ± 0.27 μm). Clear seasonal variations of the mode μ can be seen. Mostly, the variations of the mode μ are not in phase with the mass, the largest particles are mainly found shortly before or behind the dust peak. Mode μ and logarithmic mass show no correlation (r = 0.13 at the 90 % significance level) in the LGM, whereas a weak correlation is found in the late glacial (r = 0.38 at the 99.5 % significance level).

5.3.4 Discussion and Conclusion

A pronounced seasonal dust cycle was observed for the early Holocene and for the first time also in the last glacial period. A sharp increase to the dust maximum, especially in the Allerød IS, LGM and late glacial period might be attributable to an abrupt modification of the atmospheric circulation in the source areas where the seasonality is influenced by the migration of the Asiatic polar front (Roe 2009). At present day the dust peaks found in the ice occur concurrent with dust storms in the Taklamakan desert, mainly observed in spring (Bory et al. 2002). In the Holocene and Allerød IS a secondary attenuated dust season exists in the course of the year, which might be caused either by precipitation events or by transport processes. Due to similar dust patterns in the presented periods we suggest that also in the glacial the dusty season occurred in spring. The seasonal variation of mode and mass concentration is mostly not in phase, although slightly coarser particles were found in spring, which might be an indicator for paleo-storminess activity in the

Asian deserts and/or direct transport routes towards the interior of Greenland (e.g. Zielinski et al. 1997). The finer particles (<2 μm) obtained in the low-dust periods (summer to winter) might be an indicator for the background dust level in the northern Hemisphere or according to Bory et al. (2002) originate from the Mongolian Gobi during that period of time.

We identified increased seasonal amplitude in the dust concentration during colder periods, which should be taken into account when interpreting long-term variability of dust concentration found in the glacial/interglacial transition.

Acknowledgments This work was supported by funding from MARUM. The North-GRIP project was directed and organized by the Department of Geophysics at the Niels Bohr Institute for Astronomy, Physics and Geophysics, University of Copenhagen. It was supported by funding agencies in Denmark(SNF), Belgium (NFSR), France (IFRTP and INSU/CNRS), Germany (AWI), Iceland (RannIs), Japan (MECS), Sweden (SPRS), Switzerland (SNF) and the United States of America (NSF). We wish to thank all the funding bodies and field participants.

References

Archer D, Maier-Reimer E (1994) Effect of deep-sea sedimentary calcite preservation on atmospheric CO_2 concentration. Nature 376:260–263

Biscaye PE, Grousset FE, Revel M, Van der Gaast S, Zielinski GA, Vaars A, Kukla G (1997) Asian provenance of glacial dust (stage 2) in the Greenland ice sheet project 2 ice core, summit. Greenland J Geophys Res 102:26765–26781

Bory AM, Biscaye P, Svensson A, Grousset F (2002) Seasonal variability in the origin of recent atmospheric mineral dust at NorthGRIP. Greenland Earth Plan Sci Lett 196:123–134. doi:10.1016/S0012-821X(01)00609-4

Broecker WS, Peng T-H (1987) The role of $CaCO_3$ compensation in the glacial to interglacial atmospheric CO_2 change. Glob Biogeochem Cycles 1:15–29

Emerson SR, Hedges JI (2008) Chemical oceanography and the marine carbon cycle. Cambridge University Press, New York

Fischer H, Schmitt J, Lüthi D, Stocker TF, Tschumi T, Parekh P, Joos F, Köhler P, Völker C, Gersonde R, Barbante C, Floch ML, Raynaud D, Wolff E (2010) The role of Southern ocean processes on orbital and millennial CO_2 variations—a synthesis. Quat Sci Rev 29:193–205

Foster LC, Finch AA, Allison N, Andersson C, Clarke LJ (2008) Mg in aragonitic bivalve shells: seasonal variations and mode of incorporation in *Arctica Islandica*. Chem Geol 254:113–119. doi:10.1016/j.chemgeo.2008.06.007

Gaffey SJ, Bronnimann CE (1993) Effects of bleaching on organic and mineral phases in biogenic carbonates. J Sediment Res 63:752–754

Harrison SP, Kohfeld KE, Roelandt C, Claquin T (2001) The role of dust in climate changes today, at the last glacial maximum and in the future. Earth-Sci Rev 54(1–3):43–80. doi:10.1016/S0012-8252(01)00041-1

Hodell DA, Venz KA, Charles CD, Ninnemann US (2003) Pleistocene vertical carbon isotope and carbonate gradients in the South Atlantic sector of the Southern ocean. G-cubed. doi:10.1029/.2002GC000367

Keir RS (1988) On the late pleistocene ocean geochemistry and circulation. Paleoceanography 3:413–445

Key RM, Kozyr A, Sabine CL, Lee K, Wanninkhof R, Bullister JL, Feely RA, Millero FJ, Mordy C, Peng T-H (2004) A global ocean carbon climatology: results from global data analysis project (GLODAP). Glob Biogeochem Cycles. doi:10.1029/2004GB002247

Köhler P, Fischer H, Munhoven G, Zeebe RE (2005) Quantitative interpretation of atmospheric carbon records over the last glacial termination. Glob Biogeochem Cycles. doi:10.1029/2004GB002345

Krause-Nehring J, Klügel A, Nehrke G, Brellochs B, Brey T (2011a) Impact of sample pretreatment on the measured element concentrations in the bivalve Arctica islandica. Geochem Geophys Geosyst 12:Q07015. doi:10.1029/2011GC003630

Krause-Nehring J, Thorrold SR, Brey T (2011b) Trace element ratios (Ba/Ca and Mn/Ca) in Arctica islandica shells—is there a clear relationship to pelagic primary production? J Geophys Res-Biogeo (submitted)

Krause-Nehring J, Thorrold SR, Brey T (2012) Centennial records of lead contamination in northern Atlantic bivalves (Arctica islandica). Mar Pollut Bull 64:233–240. doi: 10.1016/j.marpolbul.2011.11.028

Love KM, Woronow A (1991) Chemical changes induced in aragonite using treatments for the destruction of organic material. Chem Geol 93:291–301. doi:10.1016/0009-2541(91)90119-C

Mahowald NM, Yoshioka M, Collins WD, Conley AJ, Fillmore DW, Coleman DB (2006) Climate response and radiative forcing from mineral aerosols during the last glacial maximum, pre-industrial, current and doubled-carbon dioxide climates. Geophys Res Lett 33. doi:10.1029/2006GL026126

Marchitto TM, Lynch-Stieglitz J, Hemming SR (2005) Deep Pacific $CaCO_3$ compensation and glacial-interglacial atmospheric CO_2. Earth Planet Sci Lett 231:317–336

Monnin E (2006) EPICA dome C high resolution carbon dioxide concentrations. doi:10.1594/PANGAEA.47248

Nriagu JO (1990) The rise and fall of leaded gasoline. Sci Total Environ 92:13–28. doi:10.1016/0048-9697(1090)90318-O

Orsi AH, Whitworth T, Nowlin WD (1995) On the meridional extent and fronts of the Antarctic circumpolar current. Deep Sea Res Pt I(42):641–673

Pierrot DEL, Wallace DWR (2006) MS excel program developed for CO_2 system calculations. ORNL/CDIAC-105a. Carbon dioxide information analysis center, Oak Ridge National Laboratory, U.S. Department of Energy, Oak Ridge, Tennessee. doi:10.3334/CDIAC/otg.CO2SYS_XLS_CDIAC105a

Raitzsch M, Hathorne EC, Kuhnert H, Groeneveld J, Bickert T (2011) Modern and late pleistocene B/Ca ratios of the benthic foraminifer planulina wuellerstorfi determined with laser ablation ICP-MS. Geology 39:1039–1042

Roe G (2009) On the interpretation of Chinese loess as a paleoclimate indicator. Quat Res 71:150–161

Ruth U, Wagenbach D, Steffensen JP, Bigler M (2003) Continuous record of microparticle concentration and size distribution in the central Greenland NGRIP ice core during the last glacial period. J Geophys Res 108:4098–4110. doi:10.1029/2002JD002376

Schöne BR, Houk SD, Castro ADF, Fiebig J, Oschmann W, Kröncke I, Dreyer W, Gosselck F (2005) Daily growth rates in shells of Arctica islandica: assessing sub-seasonal environmental controls on a long-lived Bivalve Mollusk. Palaios 20:78–92. doi:10.2110/palo.2003.p2103-2101

Sigman DM, Hain MP, Haug GH (2010) The polar ocean and glacial cycles in atmospheric CO_2 concentration. Nature 466:47–55

Stecher HA, Krantz DE, Lord CJ, Luther GW, Bock KW (1996) Profiles of strontium and barium in Mercenaria mercenaria and Spisula solidissima shells. Geochim Cosmochim Ac 60:3445–3456. doi:3410.1016/0016-7037(3496)00179-00172

Steffensen JP (1997) The size distribution of microparticles from selected segments of the Greenland ice core project ice core representing different climatic periods. J Geophys Res 102(C12):26755–26764

Steffensen JP, Andersen KK, Bigler M, Clausen HB, Dahl-Jensen D, Fischer H, Goto-Azuma K, Hansson M, Johnsen SJ, Jouzel J, Masson-Delmotte V, Popp T, Rasmussen SO, Röthlisberger R, Ruth U, Stauffer B, Siggaard-Andersen ML, Svensson A, White JWC (2008) High-resolution Greenland ice core data show abrupt climate change happens in few years. Science 321(5889):680–684. doi:10.1126/science.1157707

Wiltshire KH, Dürselen C-D (2004) Revision and quality analyses of the Helgoland Reede long-term phytoplankton data archive. Helgoland Mar Res 58:252–268. doi:210.1007/s10152-10004-10192-10154

Yu J, Elderfield H (2007) Benthic foraminiferal B/Ca ratios reflect deep water carbonate saturation state. Earth Planet Sci Lett 258:73–86

Zielinski GA, Mershon GR (1997) Paleoenvironmental implications of the insoluble micropar-ticle record in the GISP2 (Greenland) ice core during the rapidly changing climate of the pleistocene-holocene transition. Geol Soc Am Bull 109:547–559

Chapter 6
Ecosystems and Climate Change

6.1 Predicting Habitat Suitability for Cold-Water Coral *Lophelia pertusa* Using Multiscale Terrain Variables

Ruiju Tong (✉), Autun Purser and Vikram Unnithan
Jacobs University Bremen, Germany
e-mail: v.unnithan@jacobs-university.de

Abstract *Lophelia pertusa* is the most common reef framework-forming cold-water coral species. The complex reef structure is known to support a high diversity of benthic species. Mapping *L. pertusa* distribution is essential for resource management, but challenging given the remoteness of their habitats. In this study, maximum entropy modelling (Maxent) was used to predict the potential distribution of *L. pertusa* at the Traena Reef on the Norwegian margin, with multiscale (30, 50 and 90 m) terrain variables being used in the model run. Maxent successfully predicted the potential distribution of *L. pertusa* at the Traena Reef. The suitable habitat was predicted to occur on the easterly tips of extended topographic features. Jackknife tests showed the terrain variables slope, aspect and plan curvature (at scale 50 m) were the most useful terrain parameters for habitat prediction of *L. pertusa* when used in isolation. The live *L. pertusa* occurrence at the Traena Reef is to a large degree influenced by local scale terrain features, with elevated areas of extant reef structures facing into prevalent current flows being most suitable for ongoing *L. pertusa* growth and reef development.

Keywords *Lophelia pertusa* · Cold-water coral · Habitat suitability modelling · Maxent · Multiscale terrain variables · Norwegian margin

G. Lohmann et al. (eds.), *Earth System Science: Bridging the Gaps between Disciplines*, SpringerBriefs in Earth System Sciences, DOI: 10.1007/978-3-642-32235-8_6, © The Author(s) 2013

6.1.1 Introduction

The cold-water coral (CWC) *Lophelia pertusa* (azooxanthellate scleractinian) has a reported wide distribution, with highest densities observed to date occurring from the southwestern Barents Sea along the eastern Atlantic continental margin southward to West Africa (Roberts et al. 2009). As the most common reef framework-forming CWC species, *L. pertusa* reefs are known to support a high diversity of benthic species (>1,300 species, in the Northeast Atlantic; Roberts et al. 2009). Mapping *L. pertusa* distribution is fundamental for assessing possible anthropogenic impacts on these islands of seafloor biodiversity, and for development of conservation management plans. Such endeavors however, are complex, given the remoteness of their habitats. Modelling habitat suitability for *L. pertusa* may give an indication of their potential distribution.

Cold-water corals are generally found on topographic relief structures (Fosså et al. 2005; Mortensen and Buhl-Mortensen 2004). The elevated seabed topography may influence the coral distribution by governing the local current regimes, thus enhancing food particle delivery and larvae concentration (Mortensen and Buhl-Mortensen 2004).

The use of terrain variables, comprising the four types, slope, aspect, curvature and bathymetric position index (BPI), as well as terrain complexity, were comprehensively summarized by Wilson et al. (2007). Multiscale terrain analysis can be used to describe seabed topographic features across a range of scales that may provide distinct habitats, thereby supporting particular species (Wilson et al. 2007). The multiscale terrain variable approach has been applied to model the potential distribution of scleractinian CWCs successfully on the Irish margin, previously (Dolan et al. 2008; Guinan et al. 2009).

Maximun entropy modelling (Maxent) is a newly-developed predictive modelling technique, based on the maximum entropy theory (Phillips et al. 2006). Maxent has proved to outperform the Genetic Algorithm for Rule-Set Production (Phillips et al. 2006) and the Ecological Niche Factor Analysis methods (Tittensor et al. 2009).

In this study we employed the Maxent model to predict habitat suitability for *L. pertusa* across a region of the Traena Reef complex based on multiscale terrain variables.

(1) Can Maxent predict the potential distribution of *L. pertusa* at local scales using terrain variables at Traena Reef?
(2) How is the potential distribution of *L. pertusa* predicted at Traena Reef?
(3) In what way does local topography at Traena Reef appear to influence *L. pertusa* distribution?

Fig. 6.1 Overview map with the study area—Traena Reef on Mid-Norwegian continental shelf

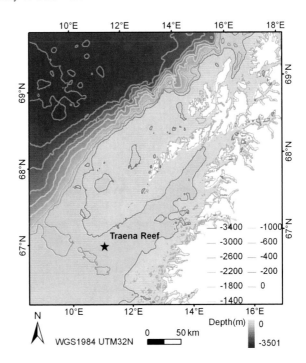

6.1.2 Materials and Method

6.1.2.1 Study Area

The Traena Reef complex is located in a sheltered embayment on the northern edge of the Traenajupet trough on the Norwegian continental shelf (Fig. 6.1). A large number of reefs of similar size have been observed at Traena Reef, with these reefs elongated in WSW–ENE direction for ~150 m (Fosså et al. 2005).

In this study, ship-borne multibeam bathymetry (Fig. 6.2, surveyed by Institute of Marine Research, Norway in 2003), videos (obtained from manned submersible JAGO dives) with associated positioning data (surveyed during Polarstern ARK XXII/1a Expedition, 2007) were used to model the potential distribution of *L. pertusa*. Still images were extracted with a 5 s time-interval from three JAGO video transects, with occurences of living *L. pertusa* logged using the software packages Adelie 1.8 and ArcGIS 9.2.

6.1.2.2 Multiscale Terrain Variables

The slope, aspect, mean curvature, plan curvature, profile curvature and (BPI) were calculated at analysis scales 30, 50 and 90 m from the bathymetry data, corresponding to moving window sizes of 3 × 3, 5 × 5 and 9 × 9 (Dolan et al. 2008;

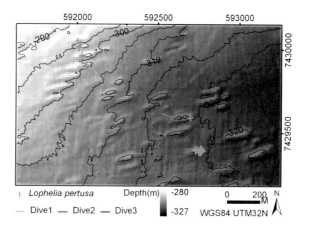

Fig. 6.2 Bathymetry data of the study area. *The lines* are the transect sections of the three JAGO dives with useful video data

Guinan et al. 2009; Wilson et al. 2007). Additionally, the terrain variables of rugosity and terrain ruggedness index (TRI) were calculated at the analysis scale of 30 m. In total, 20 multiscale terrain variables, along with depth, were used to model potential distribution of *L. pertusa*.

6.1.2.3 Predictive Method: Maxent

Maxent is a general-purpose machine learning method, which predicts a species probability distribution subject to a set of known constraints (linear features, hinge features, etc.) of all environmental variables on species distribution (Phillips and Dudik 2008). The hinge features model arbitrary piecewise linear responses to the environmental variables (Phillips and Dudik 2008). The 'target-group' background sampling data has proved to be preferable than random background sampling as its utilization largely decreases the influence of sampling bias (Phillips and Dudik 2008).

In this study, the default model settings were used, with the exception that hinge features were employed alone and that the 'target-group' background data was applied. *L. pertusa* occurrence data was randomly split into two partitions, 80 % for training and the remaining 20 % for testing with the AUC (area under ROC (receiver operating characteristic) curve) value calculated to evaluate the model performance. Jackknife tests were performed to measure the importance of terrain variables in prediction when used in isolation.

6.1.3 Results

The model evaluation showed the training AUC value to be 0.958, and the test AUC value to be 0.956. The model performed significantly better than random ($P < 0.0001$, Wilcoxon rank-sum test).

Fig. 6.3 Predicted habitat
suitability for *L. pertusa* at
Traena Reef. The high values
indicate the area more
suitable for *L. pertusa*
distribution

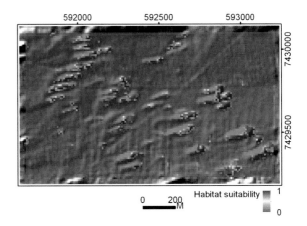

The suitable habitat of *L. pertusa* (Fig. 6.3) was predicted to occur predominantly on the easterly tips of the reefs, and along some of the reefs in the NW corner of the study site.

Jackknife tests indicated (1) plan curvature (at analysis scales of 50 and 90 m), and aspect and slope (at scale 50 m) to contain most useful information for predicting distribution of *L. pertusa* when used in isolation; (2) no one terrain variable contained a substantial amount of useful information that was not already contained to some degree within other variables for *L. pertusa* distribution modelling.

6.1.4 Discussion and Conclusion

The suitable habitat of *L. pertusa* (Fig. 6.3) was predicted to occur predominantly on the easterly tips of the reef structures, which is consistent with the observations from the JAGO dives conducted in this study. Such growth morphology of coral reefs exposed to stable, prevalent currents has been reported (Buhl-Mortensen et al. 2010). This unidirectional reef development may well be related to the prevalent bottom current flow direction (Fosså et al. 2005), with the successive generation of *L. pertusa* polyps growing most successfully in locations exposed to the greatest flux of food, in this case the upstream side of the existing reef structures.

The jackknife tests highlighted that slope, aspect and plan curvature at analysis scale 50 m contained the most useful information in predicting distribution when used in isolation. Slope may act as a proxy for suitable substrate type or for regularly occurring suitable current conditions for maximum prey capture (Dolan et al. 2008). In this study, *L. pertusa* was observed mostly on positions of topographic divergences and/or steep slopes. The aspect variable indicates orientation of an area of the seabed (Wilson et al. 2007), and in locations such as at the Traena Reef complex with a prevalent current flow direction, this variable therefore also indicates which areas are exposed to the greatest fluxes of incoming food material. Locations with such an aspect, when combined with elevated topographic height and a suitable flow velocity regime for successful prey capture and feeding (Purser et al. 2010) appear from the Maxent predictions and JAGO observations to be the most suitable for vigorous *L. pertusa* growth at the Traena Reef. Additionally, these terrain parameters at scale 50 m with most importance in prediction when used in isolation, which effectively captured the terrain variation of reef features, may suggest that local terrain features, here the reef structures, play significant roles in determining *L. pertusa* distribution and reef development.

References

Buhl-Mortensen L, Vanreusel A, Gooday AJ, Levin LA, Priede IG, Buhl-Mortensen P, Gheerardyn H, King NJ, Raes M (2010) Biological structures as a source of habitat heterogeneity and biodiversity on the deep ocean margins. Mar Ecol 31(1):21–50. doi:10.1111/j.1439-0485.2010.00359.x

Dolan MFJ, Grehan AJ, Guinan JC, Brown C (2008) Modelling the local distribution of cold-water corals in relation to bathymetric variables: adding spatial context to deep-sea video data. Deep-Sea Res PT I 55(11):1564–1579

Fosså JH, Lindberg B, Christensen O, Lundälv T, Svellingen I, Mortensen PB, Alvsvåg J (2005) Mapping of *Lophelia* reefs in Norway: experiences and survey methods. In: Freiwald A, Roberts JM (eds) Cold-water corals and ecosystems. Erlangen Earth conference series. Springer, Berlin, pp 359–391

Guinan J, Brown C, Dolan MFJ, Grehan AJ (2009) Ecological niche modelling of the distribution of cold-water coral habitat using underwater remote sensing data. Ecol Inform 4(2):83–92

Mortensen PB, Buhl-Mortensen L (2004) Distribution of deep-water gorgonian corals in relation to benthic habitat features in the northeast channel (Atlantic Canada). Mar Biol 144(6):1223–1238. doi:10.1007/s00227-003-1280-8

Phillips SJ, Dudik M (2008) Modeling of species distributions with Maxent: new extensions and a comprehensive evaluation. Ecography 31(2):161–175

Phillips SJ, Anderson RP, Schapire RE (2006) Maximum entropy modeling of species geographic distributions. Ecol Model 190(3–4):231–259

Purser A, Larsson AI, Thomsen L, van Oevelen D (2010) The influence of flow velocity and food concentration on *Lophelia pertusa* (Scleractinia) zooplankton capture rates. J Exp Mar Biol Ecol 395(1–2):55–62

Roberts JM, Wheeler A, Freiwald A, Cairns S (2009) Cold-water corals: the biology and geology of deep-sea coral habitats. Cambridge University Press, Cambridge

Tittensor DP, Baco AR, Brewin PE, Clark MR, Consalvey M, Hall-Spencer J, Rowden AA, Schlacher T, Stocks KI, Rogers AD (2009) Predicting global habitat suitability for stony corals on seamounts. J Biogeogr 36(6):1111–1128

Wilson MFJ, O'Connell B, Brown C, Guinan JC, Grehan AJ (2007) Multiscale terrain analysis of multibeam bathymetry data for habitat mapping on the continental slope. Mar Geod 30(1):3–35

Chapter 7
Geoinformatics

7.1 Resource-Aware Decomposition and Orchestration of Geoprocessing Requests in a SOA Framework

Michael Owonibi (✉) and Peter Baumann
Jacobs University Bremen, Bremen, Germany
e-mail: p.baumann@jacobs-university.de

Abstract Current trends specify the use of service-oriented, distributed computing for executing geoprocessing tasks. This is motivated by the availability of high speed networks, distributed data sets, high computation cost of geoprocessing tasks, increase in geoprocessing application requirements, server capabilities and limitations, and several more. However, it turns out that it is difficult to find a means of dynamically composing services efficiently using the resource characteristics of a set of heterogeneous servers. Therefore, in addressing these issues with respect to multi-dimensional raster data processing, we propose a query based approach using Open Geospatial Consortium (OGC) Web Coverage Processing Service (WCPS) standard. Hence, a server, after receiving a WCPS query request, can automatically extract portions of the query that are best resolved locally, distribute other requested parts to other suitable servers, re-collect results and package them. Some of the main contributions of our work include a framework where several servers can share computation, load and data, a synchronized and mirrored service registry infrastructure, server calibration methodology, coverage processing cost model, distributed query optimization and scheduling algorithms, and a P2P-based orchestration model for executing the distributed query.

Keywords Geo-processing · WCPS · Service orchestration · Service registry · Query optimization · Scheduling

G. Lohmann et al. (eds.), *Earth System Science: Bridging the Gaps between Disciplines*, SpringerBriefs in Earth System Sciences,
DOI: 10.1007/978-3-642-32235-8_7, © The Author(s) 2013

7.1.1 *Introduction*

Service Oriented Architecture (SOA) is often used today for web service framework where distributed geo-processing tasks are required. Distributed geo-processing is necessitated by the distributed original locations, size, and costly evaluation algorithms of geo-data. Such distributed processes should be embedded into some overall service which hides the complex service workflow orchestration tasks behind simple, easy-to-use interfaces. However, it turns out that it is difficult to find a means of dynamically composing and orchestrating services efficiently. We claim that this is due to the lack of semantic description of the data and services. Therefore, the orchestrations of geoprocessing services are, typically, performed manually using hardwired processing configurations, such as in the special case of cascading OGC Web Map Service (WMS) requests, Web Processing Service (WPS) chaining, WPS-Transactional and Business Process Execution Language (BPEL) based OGC services orchestration. Conversely, the semantic web facilitates the automated discovery, composition and execution of the web service, however, it is difficult to establish not just syntax (i.e., function signatures) but semantics (i.e., effects), and besides it is not efficient in composing services.

Therefore, in addressing these issues with respect to multi-dimensional raster data processing, we propose a query based approach which uses OGC Web Coverage Processing Service (WCPS) standard. WCPS specifies the syntax and semantics of a query language which allows for server-side retrieval and processing of multi-dimensional geospatial coverages (Baumann 2008). Coverages, a type of geospatial data, encompass any spatio-temporally extended phenomenon. As currently overarching query languages in this generality are not sufficiently understood, WCPS focuses on multi-dimensional raster data. Common examples include satellite imagery, thematic maps, digital elevation models etc. WCPS queries are given as expressions composed from coverage-centric primitives. Such primitives include geometric, algebraic-based, summarization, format encoding, reprojection, etc. operations. As such, the WCPS language can be understood as a declarative notation for a coverage processing workflow. Therefore, WCPS can be viewed as a process tree as shown in Fig. 7.1 where OP_x represents coverage processing operators. Due to the fact that the semantics of the service request is known both to the client and servers, Owonibi and Baumann (2009) presume that WCPS servers have the potentials for decomposing, dispatching, chaining and optimizing services automatically without any human interference.

7.1.2 *Query Decomposition and Orchestration*

In this paper, we propose a framework where several coverage processing servers can efficiently and dynamically share load and computations specified as a declarative query. A similar framework has been used in distributed databases and

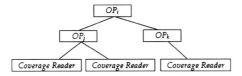

Fig. 7.1 Tree representation of WCPS

grid computing. However, their algorithms are designed for relational and xml databases as opposed to a coverage database. Besides, these use a mediator-based approach, whereby a central server parses, optimizes partitions, and schedules a query; centrally coordinate the execution of a query schedule, and integrates partial query results. Therefore, the mediator constitutes performance bottleneck and single point of failure. Furthermore, the overhead of running the grid framework significantly reduces the efficiency of distributed processing of coverages.

In the infrastructure we designed, every server can act as the mediator. Hence, each server can schedule and orchestrate a distributed query. The components of each server are shown in Fig. 7.2.

After receiving a query, the server parses and optimizes the query. The optimizations are broadly divided into two—single node and multi-node optimizations. Single-node optimization re-arranges the ordering of operators of a query tree for a efficient execution on a single server. Overall, the idea is to minimize the size of the data processed by an operator. This is because smaller input data for an operator implies less processing work to be done by an operator and smaller data transfer time between an operator and its operand. Examples of single node optimizations include pushing down geometric operations on the query tree, and the re-shuffling of operators to find the optimal operators arrangement using their commutative and associative properties. On the other hand, multi-node algorithms re-arrange the order of execution of operators to create bushy trees from left-deep trees (in order to obtain better parallelism for the operators) as shown in Fig. 7.3. It also brings operators which processes data on the same host as close together as they can possibly get.

Furthermore, in order to be able to determine the cost of processing of each operator on different node, we provide an algorithm for the modelling the cost of executing of operators (Owonibi and Baumann 2010b). The components of the cost function include the data transfer, read and write cost, and CPU cost. Also, since it is difficult to base the performance of a system on its quoted speed, we provided a calibrator for calibrating each system with respect to coverage processing. Calibrators publish their execution result in the service registry.

After the optimization, we use a modified form of Heterogeneous Earliest Finish Time (HEFT) algorithm (Topcuoglu et al. 1999) to schedule the coverage processing operators. Some of the improvements we made include—enabling intra-operator parallelism of expensive operators by data partitioning, clustering of

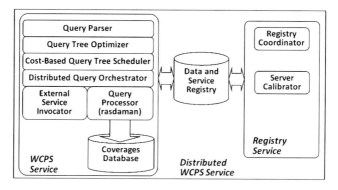

Fig. 7.2 Distributed WCPS server components

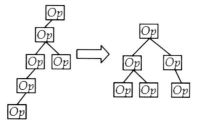

Fig. 7.3 Query tree transformation

the operators, incorporating the fact that one server can process multiple operations simultaneously without significant change in performance, and many more.

Finally, we modified the query structure such that distributed processing
information can be included. The distributed query schedule is, then, executed
using a Peer-to-Peer (P2P) paradigm (Owonibi and Baumann 2010a), whereby
each server recursively invoke other servers with distributed WCPS queries. Once
the distributed query is received, the server will invoke other servers with queries
(specified in the query it receives), execute its local query, merge the result and
ship it back to its client (the host which sent distributed query).

A central component of our proposed infrastructure is the registry. It stores the
server capabilities of all servers in the network together with their data properties.
All servers keep a local copy of the registry, and this is synchronized across all
servers in the network. Hence, every change e.g. addition and removal of servers
or data, change in server and data properties are monitored and propagated to all
servers in the network efficiently. In order to achieve this, we structure the network
such that each server has only one gateway to the network; however, it can serve as
gateway for many other servers which it monitors as shown in Fig. 7.4.

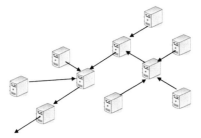

Fig. 7.4 Mirrored registry synchronization

7.1.3 *Performance Evaluation*

The performance of the execution of a query depends on the structure of the query, heterogeneity of servers, and the number of servers available. Some of the query properties include maximum parallelism (maximum total number of operations parallelizable), cost-to-communication ratio (ratio of CPU cost to data transfer cost), sequential factor (ratio of cost of parallelizable to non-parallelizable operations), and skewness factor (ratio of cost of two parallelizable operations). Using these factors, we determine the parallelizability of a query. Experimentally, we determine the speedup (ratio of the time of distributed execution of a query to the time it takes on one server) of different random query structures, Overall, we obtain speed up the parallelizability of the query.

7.1.4 *Conclusion*

This paper presents an infrastructure for distributed coverages processing. Servers can join this network by registering with any server on the network after it had graduated itself against a set of server calibration parameters we proposed. This calibration info, together with metadata for the coverages available on every server is replicated on all servers. Therefore, we provide the means to efficiently synchronize this information on all the servers' registry. Based on the information in the registry, we also propose a model for estimating the costs of distributed coverages processing, query tree optimization methods, efficient query scheduling algorithms, and a P2P-based distributed query execution technique. Several servers can, therefore, efficiently collaborate in sharing data, computation, loads and other tasks with respect to dynamic, resource-aware coverages processing.

7.2 A Specification-Based Quality Model to Improve Confidence in Web Services of Multidisciplinary Earth System Science

Jinsongdi Yu (✉) **and Peter Baumann**
Jacobs University Bremen, Germany
e-mail: j.yu@jacobs-university.de

Abstract To ensure better global sharing of geospatial information in a hetero-geneous world of multidisciplinary Earth system science, quite a few web services of earth phenomena, are implemented to claim conformance to the unanimous geo standards, such as the Open GeoSpatial Consortium (OGC), in collaboration with the ISO, both of which are currently the main driving forces in open standards for interoperable geo services. Conformance assessment, directly or indirectly deter-mines that a process, product, or service meets relevant technical standards. The assessment approaches are differentiated in the domain specific standards. How-ever, how to establish a model to evaluate the provided standardization solution, and to evaluate the extent to which the specification itself supports the improve-ment of confidence in service implementations, tests and test results, is still immature in the research of geo-service interoperability. To address these issues, we derived a specification-based quality model based on the domain specific standards. Corresponding metrics are derived to quantitatively measure these qualities during the testing stage.

Keywords Geo service · Interoperability · Standardization · Conformance · Quality model · Maturity · Validity

7.2.1 Introduction

Geo raster Web services, such as satellite image services, digital aerial photo services, digital elevation data services, provide access to detailed and rich sets of geospatial information being used in the multidisciplinary field of Earth system science research, such as in atmosphere, ocean, cryosphere, solid earth, and biosphere research. The use of internet and related web technologies for accessing such geo-spatial data, as well as for performing basic queries and processing, allow us to find, share, combine, and process geo-spatial information more easily. For example, selected geological, remote sensing and climate data for island effects research could become more accessible and more easily combined within an open interoperable framework with unanimous service interfaces. But the heterogeneity of the services on the multi-source data and the corresponding processing constitute a barrier to such geo-spatial data sharing and interoperability. Open GeoSpatial Consortium (OGC), in collaboration with the ISO, develops and maintains a family of open and interoperable geo service standards. The corresponding conformance testing programmes have been set up to ensure interoperability of implementations that claim to adhere to the specifications. For example, an important, though not very visible activity in the OGC OWS-8 Observation Fusion Thread consists of establishing systematic conformance tests for the Web Coverage Service. The goal of the testing is to check that all the different venders in the network communicate in common Earth Observation service interfaces and data models. While such testing poses challenges in itself, the extent to which the specifications themselves support testing of implementations is also of significance.

7.2.2 Methods

A geo service conformance assessment process claims the conformance level to the specified geo standards during the testing stage. The external metrics (ISO 2003), which are measured during such a process, quantitatively reflect these levels. Therefore, we derived the external metrics to measure the three derived qualities, specifically, functional compliance, which is used to measure the compliance items of the service interfaces, test coverage, which is used to measure the maturity of the designed test cases, and audit trail capability, which is used to measured the analyzability of test results. These metrics are detailed in Table 7.1.

To test the compliance items, a set of test cases are designed for the tasks. However, the designed test cases can not 100 % guarantee the implementations due to the incomplete nature of testing (Dijkstra 1972). In this case, a set of well designed test cases will help to improve the confidence on the provided services. To measure the test maturity of the designed tests, a metric of test coverage is needed. Request Parameter Relationship Analysis (RPRA) (Yu et al. 2011), which helps to generate finite search spaces of different complexities by considering the specification-consistent parameter relationships, provides a measurement on the total number of test cases that are to be performed to cover a compliance item.

Test results are important to test audit ability. Obviously, simple pass or fail is inexpressive for the inconclusive verdict. Audit trail capability is used to measure analyzability based on the evaluation properly, taking into account inconclusive results. To measure the analyzability, a metric of audit trail capability is needed. We propose the Integrated Dependency and Goal Model (IDGM) to address subordinations and dependencies among the conformance statements.

7.2.3 Results

We designed the RPRA-query-language (RPRA-QL) based on the RPRA theory as specified in (Yu et al. 2011) and applied it in the OGC WCS 2.0 (Baumann 2010a, b)

Table 7.1 External metrics for specification-based services

Quality	Metrics	Formula	Parameter description
Functionality compliance	Functional compliance	A/B	A is the number of the compliance requirements as specified that have been correctly implemented during testing and B is the total number of compliance requirements specified
Maturity	Test coverage	C/D	C is the number of actually performed test cases representing operation scenario during testing the requirement and D is the number of total test cases that are to be performed to cover the requirement.
Analyzability	Audit trail capability	E/F	E is the number of data actually recorded during operation and F is the number of data planned to be recorded enough to monitor status of software during operation

test maturity analysis. We implement RPRA-QL in BNFC, a compiler construction tool generating a compiler front-end from a Labelled BNF grammar (Forsberg and Ranta 2005). An example for a RPRA-QL on a data retrieval is as follows:

IRCP(*Service*, *Version*, MDRCP(*id*, SR(MDR(d_1, SSR(TDR(*TrimL$_1$*, *TrimH$_1$*), *SlicePoint$_1$*)), MDR(d_2, SSR(TDR(*TrimL$_2$*, *TrimH$_2$*), *SlicePoint$_2$*)), ..., MDR(d_i, SSR(TDR(*TrimL$_i$*, *TrimH$_i$*), *SlicePoint$_i$*)))))

This query generates test requests, constrained by the relations among parameters based on RPRA. By comparing with static approach, random approach and exhaustive approach, the RPRA approach generates a search space which contains minimum necessary requests and these requests are specification-consistent.

In IDGM, we treat statement subordination and dependency relationships as directed edges in graph theory. We model the relationship between a statement and its direct successors in Test Result Evaluation (TRE) language, which is based on a three-valued logic (Fitting 1991) evaluation language, to evaluate validity of global statements. For example, if the truth value of Conformance Statement (CS) CS1 depends on *CS2 or CS3 and CS4 and CS5*, and CS2 consists of *CS6 and CS7 and CS8*, CS1 can be evaluated by *(CS6 and CS7 and CS8) or (CS3 and CS4 and CS5)*. An iteration process on compound statements will find all statements which are necessary to be evaluated.

7.2.4 Discussion and Conclusion

We have investigated 'testing Web services' against complex specifications, such as OGC WCS 2.0 series (Baumann 2010a, b), consisting of requirements with manifold subordinations and dependencies among them. In order to insure general applicability, a derived quality model is proposed to address service interface compliance, test maturity and global conformance statement validity. Functional compliance is used to measure the compliance items of the service interfaces. A metric of test coverage based on RPRA-QL is proposed to quantitatively evaluate the maturity of the design tests. TRE, a three-valued logic expression is modeled to evaluate the statement result validity in the proposed Integrated Dependency and Goal Model (IDGM).

We introduced three logic assumption-based evaluations to distinguish fully credible, questionable and completely not credible statements. Specifically, Open World Assumption (OWA) states that the truth value of a statement that is not included in or inferred from the knowledge explicitly recorded in the system shall be considered unknown. Closed World Assumption (CWA) is the assumption that any statement is false that is not known to be true. We distinguish yet another case in software testing. Frequently, there are test stubs which simulate the behavior of the dependent test modules. In this case, the dependent test modules are always assumed to be true when their truth evaluation results are not available. Hence, we define Stub Assumption (SA) as the presumption that what is not currently known to be true is true.

We have proven that there is one and only one fix-point in the isolated testing environment according to the Knaster–Tarski theorem (Echenique 2005). When a fixed logic assumption strategy is adopted for the truth evaluation, the fix-point location is unique. However, this fix-point location is always unknown since the logic assumptions are not always explicit during the evaluation process.

Our study example is geo service standardization, specifically geo raster services based on OGC WCS 2.0 (Baumann 2010a, b). The theoretical results described here form the basis for the generation of a concrete Executable Test Suite (ETS), which will become the official conformance test suite of WCS in OGC's Compliance and Interoperability Testing Initiative (CITE) programme.

References

Baumann P (ed.) (2008) Web coverage processing service (WCPS) language interface standard. OGC 08-068r2

Baumann P (ed) (2010a) WCS 2.0 Overview: core and extensions, OGC 19-153

Baumann P (ed) (2010b) OGC WCS 2.0 interface standard—core, OGC 19-110r3

Owonibi M, Baumann P (2009) Towards a semantic-based automatic orchestration of geo-processing services. In: 17th International conference on geoinformatics, Fairfax, USA

Owonibi M, Baumann P (2010a) Heuristic geo query decomposition and orchestration in a SOA. In: 12th International conference on information integration and web-based applications and services (iiWAS)

Owonibi M, Baumann P (2010b) A cost model for distributed coverage processing services. In: Proceedings of the ACM SIGSPATIAL international workshop on high performance and distributed geographic information systems (HPDGIS'10), ACM, New York, pp 19–26

Topcuoglu H, Hariri S, Wu Y (1999) Task scheduling algorithms for heterogeneous processors. In: 8th IEEE heterogeneous computing workshop, pp 3–14

Dijkstra EW (1972) Notes on structured programming. In: Dahl OJ, Dijkstra EW, Hoare CAR (eds) Structured programming. Academic Press, London

Echenique F (2005) A short and constructive proof of Tarski's fixed-point theorem. Int J Game Theory 33:215–218

Fitting M (1991) Kleene's logic, generalized. J Logic Comput 1(6):797–810

ISO (2003) Software engineering—product quality—part 2: external metrics, ISO/IEC TR 9126-2:2003

Forsberg M, Ranta A (2005) The user manual for BNF converter. Available at: http://www.cse.chalmers.se/alumni/markus/BNFC/LBNF-report.pdf

Yu J, Baumann P, Wang X (2011) RPRA: A Novel Approach to Mastering Geospatial Web Service Testing Complexity.; In ICSDM 252–256. DOI:10.1109/ICSDM.2011.5969042

Chapter 8
Geoengineering

8.1 Feasibility Study of Using a Petroleum Systems Modeling Software to Evaluate Basin Scale Pressure Evolution Associated With CO_2 Storage

Christian Ihrig (✉) and Vikram Unnithan

Jacobs University Bremen, Germany
e-mail: v.unnithan@jacobs-university.de

Abstract Carbon Capture and Storage (CCS) has been identified to be a promising method to reduce anthropogenic CO_2 emissions, helping to mitigate the risk of climate change. Among several challenges that come along with the industrial deployment of CCS, it is important to study and evaluate the influence of CO_2 storage project(s) on the surrounding area (i.e., outside of the reservoir). In this context, a relevant issue is the pressure evolution in the subsurface that is associated with CO_2 injection. Instead of using conventional reservoir simulators to model the pressure development, we apply the petroleum systems modeling software PetroMod® to simulate selected aspects of CO_2 storage. This software type has the advantage to include the long-term perspective of CO_2 storage, and the large spatial scale (basin wide) that is needed to fully evaluate this process. Since using a petroleum systems modelling software in the field of CO_2 storage is a new approach, we first present a method we have developed to simulate CO_2 injection in PetroMod. We then run a 2D model to test whether PetroMod can be used to model the pressure buildup resulting from CO_2 injection.

Keywords CO2 storage · CCS · Pressure · PetroMod · Basin scale · Long-term

G. Lohmann et al. (eds.), *Earth System Science: Bridging the Gaps between Disciplines*, SpringerBriefs in Earth System Sciences,
DOI: 10.1007/978-3-642-32235-8_8, © The Author(s) 2013

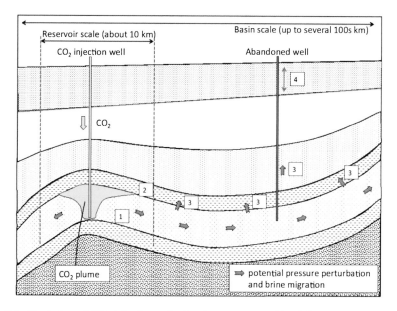

Fig. 8.1 Illustration of CO_2 injection and related processes outside of the reservoir scope (i.e., pressure perturbation and brine migration). *1* reservoir = storage formation; *2* cap rock = sealing formation; *3* potential brine leakage; *4* potential change in piezometric head due to pressure changes in the subsurface. Adapted from and modified after Birkholzer et al. (2009)

8.1.1 Introduction

Carbon Capture and Storage (CCS) is considered to be a potential method to reduce anthropogenic CO_2 emissions in order to combat climate change (IPCC 2005, 2007). If the technology is to play its part in the portfolio of CO_2 mitigation options, a large fraction of the annually released CO_2 will have to be stored in geological formations. Respective global numbers are estimated to be around 10 Gtons per annum in the year 2050 (IEA 2009). Major targets are saline aquifers, i.e., porous rock formations such as sandstones that are saturated with salty water (i.e., brine), allowing the injection of fluids like CO_2.

Apart from the premise to safely contain CO_2 below the surface (which basically includes the study of reservoir scale processes), a critical issue of industrial CCS deployment is the investigation and evaluation that CO_2 project(s) may have on the surrounding area:

During the injection period of a storage project (Fig. 8.1), the native brine is displaced by the injected CO_2, which can lead to a significant pressure buildup inside the reservoir. However, while the spreading of CO_2 is limited to the storage site (i.e., reservoir scale), the pressure front may migrate far beyond the reservoir scope, affecting a much larger subsurface volume than the CO_2 plume itself. The induced pressure change may also lead to brine migration outside of the reservoir as a result of pressure attenuation in the subsurface. Accordingly, pressure change and/or brine migration may change groundwater levels or salt contents in fresh-water reservoirs in far distance from the injection site. The relevance and risk of these issues has been reported by several recent studies (e.g., Birkholzer et al. 2009; Birkholzer and Zhou 2009; Zhou et al. 2009).

The conventional way of addressing these problems is computer modelling using reservoir modelling software types. Their advantage is their high resolution associated with great speed; however, considering the potentially affected areas of basin scale, the demanded CPU power is too high. In this context, we try to install another software tool that can deal with the relevant processes. The respective tool should be able to

(a) deal with a large number of CO_2 storage projects
(b) include important basin scale processes like the pressure-buildup and/or brine migration
(c) cover large areas and time scales up to 10,000 years to make long-term predictions.

For this approach, we use the standard industry software PetroMod®, (Schlumberger 2009) a petroleum systems modelling software, which has been designed to model geological basin history (up to several 100 million of years) with associated generation, migration and accumulation of hydrocarbons (i.e., petroleum). Accordingly, the software can cope with the considered time scale of up to 10,000 years and the large spatial scale (i.e., basin wide). However, since this type of software has never been used in the field of CO_2 storage, our goals are

(i) to develop a way to simulate CO_2 injection in PetroMod since this option is missing;
(ii) to evaluate whether we can use PetroMod to model pressure buildup resulting from CO_2 injection
(iii) to evaluate whether the software can be used to make long-term predictions concerning vertical CO_2 migration (not presented in this paper).

8.1.2 Principles of Petroleum Systems Modelling

Petroleum Systems Modelling (PSM) incorporates a large number of various processes that are needed to fully simulate and understand the evolution of geological basin with associated hydrocarbon history. Simplified, the following processes and modelling steps may be distinguished (Hantschel and Kauerauf 2009):

- Deposition, erosion and burial of sediments
- Diagenetic processes, i.e., lithology transformation (e.g., changes in porosity and permeability caused by compaction or mineral transformation); pressure calculation

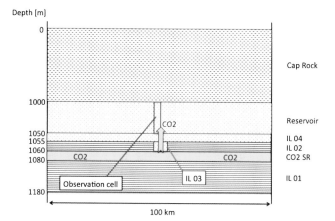

Fig. 8.2 2D model setup for the pressure evaluation in PetroMod. The figure also describes the general procedure of the developed CO_2 injection method. *IL01* impermeable layer 01; *CO_2 SR* layer where CO_2 is generated; *IL 02* impermeable layer 02; *IL 03* 2 cells that are changed from impermeable to permeable to start CO_2 "injection". *IL 04* permeable layer that can be changed to impermeable to stop CO_2 "injection". For more information, see text

- Heat flow analysis, which is needed to calculate proper subsurface temperatures; The thermal history is a crucial parameter for petroleum generation
- Fluid analysis (petroleum composition)
- Petroleum migration
- Reservoir volumetrics.

In this context, the following method should be considered as a part of a more comprehensive simulation "workflow".

8.1.3 CO_2 Injection Method in PetroMod

CO_2 has been generated in a layer (CO_2 SR, see Fig. 8.2), following the same way hydrocarbon generation is modeled in PetroMod. Two impermeable layers on the top (IL 02) and the bottom (IL 01) of the generation layer hinder the outflow of CO_2. Using a function of PetroMod (i.e., the "salt piercing tool"), a part of the upper layer (IL 03) is changed from impermeable to permeable, allowing the outflow (="injection") of CO_2 into the reservoir. This way, time and space of CO_2 "injection" can be controlled. To stop "injection", another layer (IL 04) is changed from permeable to impermeable, using the mentioned "salt piercing tool".

The presented method can be used to model one or several storage projects within one model.

8.1.4 Evaluation of the Pressure Buildup in PetroMod

8.1.4.1 Method

We choose a 2D model (Fig. 8.2) to test whether we can use PetroMod to model the pressure development following CO_2 injection. The model domain (100 × 1 × 1,18 km) is discretized into 8000 cells. Each layer consists of 1000 cells except for the Cap Rock, which is split vertically by a factor of 3 into 3000 cells due to its thickness. The dimensions of each cell in x-, y- and z-direction are 100 × 1000 m × layer thickness of the corresponding layer. The storage formation (Reservoir) is represented by a sandstone layer (porosity = 0.32, vertical permeability = 3.47D), the sealing formation (Cap Rock) by shale (porosity = 0.3, vertical permeability = 2.6 μD). We inject 4 Mtons of CO_2 over a timespan of 100 years, using the introduced injection method. The pressure response is recorded in four observation cells (of which only one is presented here) for the next 500 years, i.e., 600 years in total; the observation cell is directly located above the injection point (Fig. 8.2). To verify that the recorded pressure is a result of the injected CO_2, we also run a reference scenario where no CO_2 is injected (following the protocol of the injection method but without generating CO_2 in the CO2 SR).

8.1.4.2 Results

Hundred years after CO_2 injection, the pressure in our observation cell has increased by 0.1 bar. Throughout the rest of the simulation (i.e., another 500 years), the pressure remains constant. The reference scenario, where no CO_2 is injected, delivers identical results.

8.1.5 Discussion and Conclusion

In reality, the injection of CO_2 should lead to a significant pressure change, which is not observed in our model. Instead, the absence/presence of CO_2 does not influence our results. We are testing the model for consistency and also run a discussion on what is causing the small pressure increase in PetroMod. At the moment, we consider the observed pressure increase (0.1 bar) 100 years after injection to be a combined result of:

(a) aquathermal expension of the CO_2 SR formation water, i.e., brine that is trapped in that layer (along with or without CO_2); this leads to a significant pressure increase in CO_2 SR (around 10 bars above hydrostatic pressure, not presented here); and

(b) the pressure compensation between the Reservoir (in situ pressure is about hydrostatic, not presented here) and CO_2 SR that follows after IL 03 is opened to start CO_2 injection.

With regard to the current state of our research, i.e., the absence/presence of CO_2 in the model does not affect the pressure results, we consider PetroMod not to be suitable to study subsurface pressure evolution associated with CO_2 injection. Accordingly, we favor/support the application of reservoir simulators to evaluate this process in combination with computer techniques (e.g., parallel processing) to accelerate the model.

Acknowledgments The authors would like to thank Schlumberger GmbH for free access and use of PetroMod®, and the Helmholtz Association for funding this project. Furthermore, parts of the study were carried out at the Centre for CO_2 Storage at the Deutsche Geoforschungszentrum, Potsdam. Accordingly, the authors would like to thank Dr. Bernd Wiese, Dr. Thomas Kempka and Dr. Michael Kühn for their support and feedback during and after the study.

References

Birkholzer JT, Zhou Q (2009) Basin-scale hydrogeologic impacts of CO_2 storage: capacity and regulatory implications. Int J Greenh Gas Control 3:745–756

Birkholzer JT, Zhou Q, Tsang C-F (2009) Large-scale impact of CO_2 storage in deep saline aquifers: a sensitivity study on pressure response in stratified systems. Int J Greenh Gas Control 3:181–195

Hantschel T, Kauerauf AI (2009) Fundamentals of basin and petroleum systems modeling. Springer, Berlin

IEA (International Energy Agency): Technology Roadmap: Carbon Capture and Storage (2009) http://www.iea.org/papers/2009/CCS_Roadmap.pdf. Accessed 13 Mar 2012

IPCC: IPCC Special Report on Carbon Dioxide Capture and Storage. Prepared by Working Group III of the Intergovernmental Panel on Climate Change [B. Metz, O. Davidson, H.C. de Coninck, M. Loos, and L.A. Meyer (eds.)]. Cambridge University Press, Cambridge, United Kingdom and New York, NY, USA, 442 pp. (2005)

IPCC: Climate Change 2007: Synthesis Report. Contribution of Working Groups I, II and III to the Fourth Assessment Report of the Intergovernmental Panel on Climate Change [Core Writing Team, Pachauri, R.K. and Reisinger, A. (eds.)]. IPCC, Geneva, Switzerland, 104 pp. (2007)

Schlumberger IES PetroMod®, Petroleum Systems Modeling Software, Release 11 (2009) http://www.slb.com/services/software/geo/petromod_modeling_software.aspx. Accessed 13 Mar 2012

Zhou Q, Birkholzer JT, Mehnert E, Lin Y-F, Zhang K (2009) Modeling basin- and plume-scale processes of CO_2 storage for full-scale deployment. Ground Water 48(4):494–514

Printed by Publishers' Graphics LLC
MO20130225.19.17.34